Galápagos Marin

GW00419205

A Field Guide to
Corals and other
Radiates of Galápagos

**An illustrated guidebook to the corals,
anemones, zoanthids, black corals,
gorgonians, sea pens, and hydroids
of the Galápagos Islands.**

By
Cleveland P. Hickman, Jr.
Washington and Lee University
Lexington, Virginia, USA

PHOTOGRAPHY BY
The author unless otherwise attributed

ORIGINAL ARTWORK BY
William C. Ober, M.D.

Sugar Spring Press   Lexington, Virginia
2008

Library of Congress Control Number: 2007909741

ISBN 978-0-9664932-4-5

Printed in the United States of America

by Progress Press, Roanoke, Virginia

Sugar Spring Press

802 Sunset Drive

Lexington, Virginia 24450

USA

Contents

vi   An Introductory Note

viii   Acknowledgements

ix   The Radiates - Phylum Cnidaria

1   Corals of Galápagos

5   Reef-Building (Hermatypic) Corals

45   Non Reef-Building (Ahermatypic) Corals

62   Sea Anemones

77   Zoanthids

92   Cerianthids (Tube Anemones)

95   Black Coral

98   Ctenophores

100   Gorgonians and Sea Pens

120   Hydroids

135   Addendum to Published Field Guides

150   Glossary

153   Selected References

159   Index

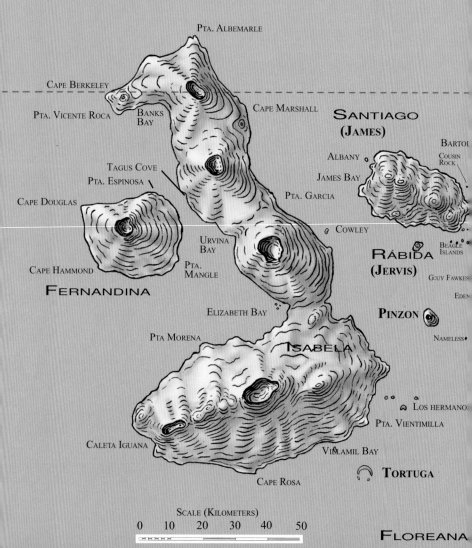

DARWIN

WOLF

PINTA

ROCA REDONDA

PTA. ALBEMARLE

CAPE BERKELEY

PTA. VICENTE ROCA   BANKS
                      BAY

CAPE MARSHALL   SANTIAGO
                  (JAMES)

BARTOL
COUSIN
ROCK

ALBANY

TAGUS COVE
PTA. ESPINOSA

JAMES BAY

CAPE DOUGLAS

PTA. GARCIA

COWLEY

URVINA
BAY

RÁBIDA
(JERVIS)

CAPE HAMMOND

PTA.
MANGLE

G:UY FAWKES

EDEN

FERNANDINA

BEAGLE
ISLANDS

ELIZABETH BAY

PINZON

PTA MORENA

ISABELA

NAMELESS

LOS HERMANO

PTA. VIENTIMILLA

CALETA IGUANA

VILLAMIL BAY

TORTUGA

CAPE ROSA

SCALE (KILOMETERS)

0   10   20   30   40   50

FLOREANA

BLACK B

THE GALÁPAGOS ISLANDS

MARCHENA

PTA. ESPEJO

GENOVESA
(TOWER)

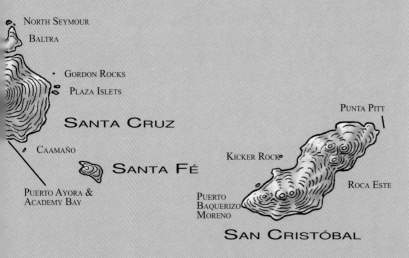

DARWIN BAY

NORTH SEYMOUR

BALTRA

GORDON ROCKS

PLAZA ISLETS

SANTA CRUZ

CAAMAÑO

SANTA FÉ

PUNTA PITT

KICKER ROCK

ROCA ESTE

PUERTO AYORA &
ACADEMY BAY

PUERTO
BAQUERIZO
MORENO

SAN CRISTÓBAL

DEVIL'S CROWN (ONSLOW ISLAND)

PTA. CORMORANT
ENDERBY & CHAMPION

CALDWELL

GARDNER

WATSON

ESPAÑOLA

GARDNER

PTA. SUAREZ

PTA. CEVALLOS

An Introductory Note

It has been eight years since the publication of the last field guides in the Galápagos Marine Life Series. It was never my intention to delay so long in bringing this long-planned guidebook to the corals and their radiate allies to publication. In researching the first three field guides to the echinoderms, marine molluscs, and crustaceans, I benefited from the extensive literature on these groups. This was not the case when I began work on the radiates. The centerpiece of the present field guide, the reef-building corals, fortunately has been, and continues to be, the subject of research publication activity by Peter Glynn and others. However, the radiate relatives of corals were poorly known. None of the publications arising from the four Allan Hancock Expeditions to the Galápagos Islands treated the anemones, zoanthids, and gorgonians, all of which are important and prominent components of the Galápagos shallow-water environment. Most of the early expeditions to the archipelago collected specimens by dredging, which resulted in far more information about the deep-water fauna than the shallow-water marine invertebrates bordering the islands. Selective inshore collecting became possible with the introduction of SCUBA in the 1950s, but research still languished on all the radiate groups except the corals.

The eight year hiatus between publication of the earlier field guides and this one enabled me to benefit from the publication of several recent and significant taxonomic studies of Galápagos marine invertebrates. Five new gorgonian species are described in recent publications by Odalisca Breedy, Gary Williams, and Hector Guzman. Odalisca Breedy subsequently identified additional gorgonians from specimens I collected in Galápagos; these are included among the 12 gorgonians and four sea pens described herein.

The identification of Galápagos sea anemones and zoanthids by researchers remains a work in progress. The inclusion of ten sea anemone and two cerianthid species in this field guide was advanced by the recent report by Daphne Fautin and coworkers. Still, several collected sea anemone specimens remain undescribed and may include new species. Work by James Reimer and John Ryland has yielded the identification of several species of Galápagos zoanthids. We applied both classical histological and molecular approaches for the zoanthids, which comprise a neglected group, especially troublesome for identification. Many of the zoanthids we have collected are almost certainly new species that remain to be described.

The hydroid fauna is well represented in Galápagos with 96 species, reflecting the excellent dispersal ability of these animals. Dale Calder's recent report lists all species previously reported from Galápagos together with 15 additional hydroids discovered since 1992. Because most hydroids are too small to be noticed despite their abundance, I have limited my coverage to the twelve species that are conspicuous.

The coverage of corals includes all 22 hermatypic (zooxanthellate) corals known from Galápagos and 16 ahermatypic (azooxanthellate) corals. Cairns (1991) listed 42 ahermatypic corals from the Galápagos, a total that will be increased by more recent discoveries. However, only 19 of these 42 species are shallow-water species, so I am confident that nearly all inshore ahermatypic corals are included in this guidebook. Because corals, especially species in the genus *Pocillopora*, offer a special challenge for identification, I have included several photographs of each species to illustrate variation in color and growth structure. On page 25 I have included a helpful table listing the principal characteristics of the 9 species of *Pocillopora* that occur in Galápagos.

Marine zoogeographers consider the Galápagos to be a unique faunal province distinct from the Panamanian Province that embraces the continental shelf from southern Mexico to approximately the Peru-Ecuador border. Endemism in Galápagos among the inshore marine invertebrates is higher than in other eastern Pacific islands, almost certainly an effect of the greater opportunities for isolation that an archipelago provides. Endemism is low for groups with good dispersal ability, such as hydroids, bryozoans, and zooxanthellate corals, but high for many other marine invertebrates with poor or limited dispersal ability. Still, endemism is strictly at the species level; there are no endemic genera among Galápagos marine invertebrates. Endemism is expectedly much more prevalent among the terrestrial flora and fauna of Galápagos because the ocean is a more formidable barrier to colonization of terrestrial life than it is to marine life.

Acknowledgements

Though preparation of this field guide would have been quite impossible without the assistance of many professional scientists who generously gave of their time to identify specimens and contribute authoritative insights from their areas of expertise. Indeed, this field guide, which has been so many years in preparation, represents their aggregate knowledge. My contribution has been to assemble this information into these pages, and errors and omissions are of course my sole responsibility. These scientists are listed below.

Hermatypic corals: Peter W. Glynn, University of Miami; Gerard M. Wellington, University of Houston
Ahermatypic corals: Stephen D. Cairns, Smithsonian Institution, Washington, D.C.
Sea anemones: Daphne G. Fautin, University of Kansas; Marymegan Daly, Ohio State University; Verena Häussermann, Huinay Scientific Field Station, Chile
Cerianthids: Tina Molodtsova, P.P. Shirshov Institute of Oceanology of the Russian Academy of Sciences, Moscow
Zoanthids: James D. Reimer, University of the Ryukyus, Okinawa; John S. Ryland, University of Wales Swansea, UK
Gorgonians and sea pens: Odalisca Breedy, Universidad de Costa Rica; Gary C. Williams, California Academy of Sciences
Hydroids: Dale R. Calder, Royal Ontario Museum, Toronto

As with the identification of specimens by specialists, I am indebted to many people for assistance in the collection and photography of marine radiates. Foremost among these is Angel Chiriboga, whose collections and photography over the past six years played a major role in the development of this book. My gratitude extends to the cooperative assistance of the personnel of the Biomarine Laboratory of the Charles Darwin Research Station, and past and present directors Rodrigo Bustamante, Graham Edgar, and Stuart Banks, who made arrangements for collecting trips and provided ongoing support. As he had with previous publications in the Galápagos Marine Life Series, Paul Humann again provided many excellent photographs of Galápagos radiates. Others who provided photographs and are acknowledged where their photographs appear are William C. Ober, Graham Edgar, Luis Vinueza, Juan-Carlos Moncayo, Sylvia Earle, Jonathan Green, Verena Häussermann, and Joshua Feingold. Although many have helped in countless ways over the years, I especially acknowledge the assistance of Angel Chiriboga, William C. Ober, Frederic Liss, and Scott Henderson, who accompanied me on numerous dive trips to assist in specimen collection and photography. William C. Ober also prepared the illustrations for the field guide.

The Radiates - Phylum Cnidaria

T he cnidarians, an ancient group numbering some 9000 species, include some of the ocean's strangest and most beautiful creatures: flowerlike sea anemones; branching, plant-like hydroids; jellyfishes, some of great size; graceful and colorful sea fans, sea whips, and sea pens that enhance submarine "gardens"; and stony corals, whose millions of years of calcareous house-building have produced great reefs and atolls. Cnidarians are widespread in shallow-water marine environments, especially in tropical or subtropical regions. Most are sessile (attached to the bottom) or, at best, slow moving.

Cnidarians are often called "radiates" because they are characterized by radial or biradial symmetry, in which the body is concentrically arranged around the oral-aboral axis. This type of primitive symmetry is ideal for both sedentary and free-floating animals, enabling them to approach the environment (or the environment to approach them) from any direction. Despite their functional simplicity, cnidarians are effective predators (all are carnivorous) that often kill and eat prey, fishes for example, that are swifter and more highly organized . All cnidarians are festooned with graceful tentacles that bristle with tiny stinging organelles, nematocysts. These discharge when touched, sending a penetrating thread bearing a paralyzing toxin into the prey. An animal unfortunate enough to brush against a tentacle is suddenly speared with hundreds or even thousands of nematocysts and quickly paralyzed. Some nematocysts are powerful enough to penetrate human skin, causing great pain, even death, depending upon the species.

A curious feature of many cnidarians is that all fit into one or both of two morphological types: a **polyp**, or hydroid form, which is adapted for a sessile, sedentary life, and a **medusa**, or jellyfish form, adapted for a free-floating existence.

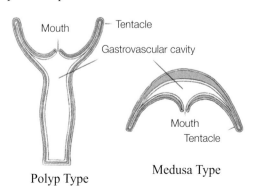

Polyp Type

Medusa Type

Attached cnidarians, such as anemones, zoanthids, gorgonians, and hydroid or coral colonies, have a polyp-type body plan. Jellyfishes, on the other hand, are of medusoid form. When we examine the two forms, we see that they are built on the same basic plan. In both, the body is a sac with a single opening, the mouth, that doubles for entrance of food and exit of indigestible wastes. The main body cavity is the intestine, or gastrovascular cavity. The colonial hydroids—delicate, feathery forms that soften rocky surfaces—often have both a polyp stage and a free-swimming medusa stage that alternate with generations—a kind of Jekyll-and-Hyde existence. In these, the hydroid stage produces medusae asexually by budding, whereas the medusae reproduce sexually, shedding eggs and sperm into the water. Fertilized eggs develop into larvae that settle to the bottom to develop into colonies consisting of polyps, completing the cycle.

The dimorphic life history is a trait unique to cnidarians; there is nothing similar anywhere in the animal kingdom. It confers great evolutionary plasticity, enabling the radiates to exploit a profusion of variations on the double-life theme.

Higher-level classification of the Phylum Cnidaria

Class Hydrozoa. Hydroids (both polyp and medusa forms)
Class Scyphozoa. Jellyfishes (medusa forms, many with polypoid larval stage)
Class Anthozoa. Corals, anemones, gorgonians and their kin (all polyp forms)
  Subclass Octocorallia (=Alcyonaria). Octocorals, with octomerous (eight-part) symmetry
    Order Gorgonacea. Gorgonians
    Order Pennatulacea. Sea pens
  Subclass Hexacorallia (=Zoantharia). Hexacorals, with hexamerous (six-part) symmetry
    Order Actiniaria. Sea anemones
    Order Scleractinia. True or stony corals
    Order Zoanthidea. Zoanthids
  Subclass Ceriantipatharia
    Order Ceriantharia. Tube anemones
    Order Antipatharia. Black or thorny corals

CORALS OF GALÁPAGOS

The true or stony corals of the order Scleractinia are metazoans that possess a hard calcareous skeleton lying outside the body of the living polyp. The polyps, resembling miniature sea anemones, have a mouth surrounded by one or more rings of tentacles that in most species can be retracted. This may be a day/night activity or a response to food or some disturbance. Adjacent polyps frequently are connected by a sheet of living tissue, the coenosarc (seen´o-sark), which is responsible for depositing skeletal material between the polyps. The skeleton of an individual polyp, the corallite, is enclosed by a wall, the theca, which consists of coenosteum (seen-os´te-um), a sponge-like skeletal matrix, and the septo-costae, radial skeletal elements of the wall. Internally, the polyp is divided longitudinally by a series of partitions, called mesenteries, that carry the gonads and are involved in digestion of food. Alternating with the mesenteries are calcareous, plate-like septa. If the septa pass over the top of the corallite to extend over the perithecal area (the surface of coenosteum between the corallites), they are called costae.

The stony corals are conveniently divided into two groups based on whether or not they build reefs. The reef-building, or **hermatypic** (from the Gr. *hermes*, reef) corals, are responsible for the very existence of a reef. Their success depends upon the presence of minute plants, called zooxanthellae, packed in their tissues. These symbiotic plants require sunlight, which explains why the most luxuriant coral reefs thrive in well-lit, relatively shallow waters. The zooxanthellae are crucially important to the hermatypic corals. Using photosynthesis to fix carbon dioxide, they furnish food molecules for their hosts, recycle phosphorus and nitrogenous waste compounds that otherwise would be lost,

The author is indebted to Peter Glynn and Stephen Cairns for assistance in the taxonomic determinations of the corals described here.

and enhance the ability of the coral to deposit calcium carbonate. Corals derive most of their nutritional requirements by active feeding, capturing drifting prey, principally plankton, on their tentacles. Many corals extend their tentacles only at night when zooplankton density in the water is greatest.

When aggregations of corals coalesce and build vertically into wave-resistant structures, the result is the formation of a **coral reef.** Historically, coral reefs in Galápagos were scattered and of small size, facts that may explain why Charles Darwin did not mention the presence of corals or coral reefs following his 5-week visit to Galápagos during the voyage of the Beagle in 1835. The deep waters surrounding Galápagos limit shallow-water habitats that corals require for reef development. Galápagos reefs also are vulnerable to periodic visits of El Niño events. In recent years Galápagos reefs were almost entirely destroyed by the 1982-83 El Niño event, which reduced coral cover by 95 to 99% (Glynn, 2003). This loss was due mainly to prolonged ocean warming, which killed the symbiotic zooxanthellae upon which the corals depend, resulting in widespread bleaching, mortality, and subsequent erosion of the coral framework. Recovery since 1983 has been very slow. The 1997-98 El Niño event, judged even more severe than the 1982-83 event, also caused bleaching of all zooxanthellate coral species, although with greater survival than after the 1982-83 event, suggesting that the coral may have adapted to some extent to El Niño conditions. Recent recovery has been more rapid than recovery following the 1982-83 event. Today, coral assemblages in Galápagos are present largely as **coral communities**. These differ from coral reefs in that new generations simply replace the old after death and erosion, rather than building vertically on existing coral structure.

The three principal genera of hermatypic corals in Galápagos are *Pocillopora, Pavona,* and *Porites.* Species of all three genera are widely distributed in Galápagos, although all flourish best on the east side of Isabela, north coasts of the central islands, and the northern islands of Darwin and Wolf (Glynn & Wellington, 1983). Substantial colonies also exist at Punta Espejo on the east side of Marchena (Banks, et al., 2008). The nine species

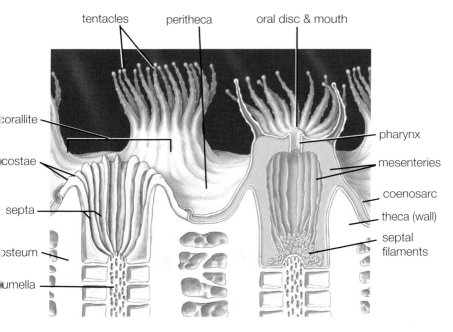

Structure of coral polyp tissues (right) and underlying corallum (left).

of *Pocillopora* included here are branching colonies with upright or prostrate branches. Branch form is important in distinguishing the species of *Pocillopora,* but consistent identification is seldom obvious and is complicated by variations in growth form due to environmental (microhabitat) effects such as wave action, current, and light availability; grazing by fish; and by hybridization among species. Hybridization occurs in corals and may lead to reticulate evolution, a form of evolution distinguished by lineages that both split (to form species) and rejoin (to form hybrids) over time (Veron, 1995, 2000). Reticulate evolution is uncommon in most animals but common in higher plants and may be widespread in corals.

  The five species of *Pavona* that occur in Galápagos form massive, columnar, or laminar colonies with corallites separated by protruding septo-costae. They tend to occur in deeper water (10-20 m) than the *Pocillopora* species (1-15 m). At the northern islands of Darwin and Wolf, *Porites lobata* is the dominant coral, often forming massive mounds. At Darwin Reef that lies between Darwin Island and Darwin's Arch, *Porities lobata* may constitute 95% of the coral present. It is a resilient species that has recovered more rapidly from El Niño events than other corals.

The two species of *Psammocora* included in this guide have highly restricted distributions in Galápagos. This genus is easily distinguished from other coral genera but the two *Psammocora* species are difficult to separate.

The **ahermatypic** (non-reef-building) corals do not contain zooxanthellae and consequently are not limited in their distribution to shallow, sunlit waters. Some thrive thousands of meters deep in the ocean. They also thrive in shallow water under ledges, in caves, even completely hidden beneath rocks, where sunlight is minimal or absent. Except for the showy *Tubastraea coccinea,* they usually are not conspicuous on reefs, although some cryptic species, such as *Culicia stellata* may be common. The corallites are rounded and cylindrical and may be completely solitary, or may be connected together at their bases by stolon-like extensions of the coenosarc to form small colonies. In many species, the polyps retract during the day. Because most are small and often inconspicuous, some ahermatypic corals are easily overlooked by divers. Other Galápagos ahermatypes are common and so vividly colored they can hardly be missed by even the most casual diver or snorkeler. These include the bright orange species *Tubastraea coccinea* and the less common *Cladopsammia eguchii*, present on vertical rock walls where they may intermingle with hermatypic corals, sponges, zoanthids, and ascidians.

REEF-BUILDING (HERMATYPIC) CORALS

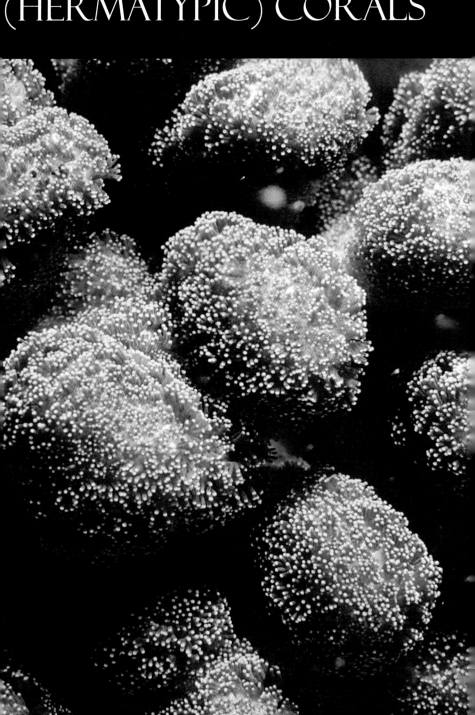

Phylum Cnidaria, Class Anthozoa, Subclass Hexacorallia,
Order Scleractinia, Suborder Astrocoeniina, Family Thamnasteriidae

Psammocora stellata Verrill, 1864. Isolated, small, branching colonies that are green or brown in color. Colonies may occur as unattached, free-moving pieces. The surface is faintly granular with the polyps often extended during the day. Calice diameter 1.28 - 2.14 mm, averaging 1.57 mm. Walls poorly defined with numerous septa that are barely visible to the unaided eye. **Habitat & range:** Shallow, wave-washed rock, or to 15-20 m depth on coarse sand bottoms in the central archipelago. Once abundant in Galápagos before the 1982-83 El Niño event, colonies now are uncommon and small, seldom forming colonies more than a few centimeters in diameter. Colonies are best observed at Devil's Crown (Onslow Island) Floreana, where they are still abundant; Gardner Bay, Española; and Pta. Espejo, Marchena. Eastern Pacific from the Gulf of California to mainland Ecuador and Galápagos, also Hawaii.

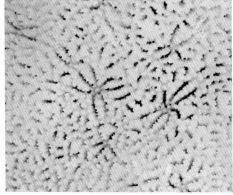

Above, Coral skeleton showing typical branching structure. x 0.6
Right, Detail of corallites. x 9

Joshua Feingold

Punta Cormorant, Floreana, 8 m

Psammocora* cf. *superficialis Gardiner, 1898. Grey or brown submassive or laminar colonies. Corallites are smaller than those of *P. stellata:* calice diameter 0.93 - 2.1 mm, averaging 1.34 mm. Polyps may extend during the day, giving the surface a furry appearance. *P. superficialis* is nearly identical to *P. profundacella,* which could represent this species as suggested by one authority (F. Benzoni, pers. comm.). Both are submassive or laminar encrusting colonies with meandering valleys. Corallite diameters are nearly identical. However, *P. profundacella* has not yet been identified in the eastern Pacific whereas *P. superficialis* is widely distributed from Mexico to Ecuador. **Habitat & range:** Rocky substrates and rubble from shallow water to 20 m. Colonies less common in Galápagos than *Psammocora stellata.* Eastern Pacific from Mexico to Ecuador and Galápagos, Hawaii, and throughout the Indo-Pacific.

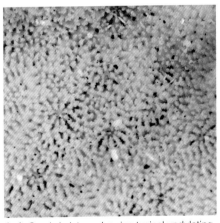

Left, Coral skeleton, showing typical undulating surface. x 0.85
Above, Closeup of coral surface showing corallites. x 9

Phylum Cnidaria, Class Anthozoa, Subclass Hexacorallia,
Order Scleractinia, Suborder Astrocoeniina, Family Pocilloporidae

Luis Vinueza

Gardner Bay, Española, 15 m

Pocillopora verrucosa (Ellis & Solander, 1786). Colonies are composed of primarily upright branches that are thick and compact in shallow water exposed to strong currents, thinner and more open in protected habitats. Branches are clearly distinct from the verrucae (va-roo'kee), which are irregular in size and usually longer than wide. This species is now considered by P. Glynn and colleagues to be conspecific with *Pocillopora elegans*. It also is similar to *Pocillopora meandrina*, which has shorter, more flattened branches and smaller verrucae (verrucae not longer than wide). **Habitat & range:** Most shallow water habitats. Widespread across the Indo-Pacific. Uncommon along mainland Ecuador.

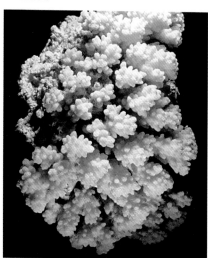

Branch tips, showing elongate verrucae.
x 0.6

Branch tip with verrucae. The presence and disposition of verrucae, wartlike mounds of coenosteum on the surface, especially distinguishes the genus *Pocillopora* . x 3.3

Pocillopora verrucosa. Note characteristic upright branches with flattened ends and prominent verrucae.

Pocillopora verrucosa. Note spreading colony and irregular spacing of verrucae.

Pocillopora verrucosa. Note thick branches with flattened ends, a characteristic of the species. The white branch tips are the result of cropping by fish, especially parrotfish, puffers, and filefish.

Graham Edgar, Wolf

Graham Edgar, Wolf

Graham Edgar, Wolf

Pocillopora damicornis (Linnaeus, 1758). Colonies are clumps of relatively thin branches that are compact in habitats with strong wave action; thin and open in protected areas. An important distinguishing character of this species is that the verrucae and branches intergrade with each other, leaving no clear distinction between the two. Cropping of tips of branches by corallivores (fishes and *Eucidaris thouarsii = galapagensis*) is common for this species of *Pocillopora*, causing misshapen tips. **Habitat & range:** Shallow water habitats. Small colonies, although uncommon, occur throughout Galápagos. This was a dominant species in the southern and central archipelago before the 1982-83 El Niño event. Gulf of California to mainland Ecuador and Galápagos; also Easter Island and throughout the Indo-Pacific.

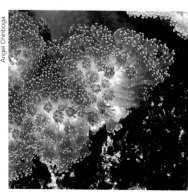

Closeup view of *P. damicornis*, showing polyps with extended tentacles.

Detail showing intergrading of verrucae and branches, a distinguishing character of this species. x 1.3

Pocillopora damicornis from Pta. Espejo, March-ena. Strong wave action in this habitat has encour-aged compact growth of this colony.

Pocillopora damicornis from Beagle Rocks, 6 m. The open branching of this colony is character-stic of growth in habitats not exposed to strong wave action.

Angel Chiriboga

Pocillopora damicornis from Darwin with extended tentacles.

Juan-Carlos Moncayo

Pocillopora meandrina Dana, 1846. Colonies are small upright bushes with flattened branches radiating from initial point of growth. Verrucae are neat and uniform, rounded and of medium size. This species is close to *Pocillopora verrucosa*, which has relatively prominent verrucae (verrucae longer than wide). Branching is compact in colonies exposed to strong current. **Habitat & range:** Shallow rocky substrates. In Galápagos distributed throughout the central archipelago and northern islands of Darwin and Wolf at 3-15 m depths. Uncommon. Widespread across Pacific, including Indo-Pacific, Hawaii, eastern Pacific from Mexico to Ecuador and Galápagos Islands.

Skeleton of a small colony showing characteristic branching. x 0.5

Branch surface showing uniform spacing of verrucae. x 3.6

Pocillopora meandrina.
A colony from Wolf, 9 m.

William C. Ober

Pocillopora meandrina.
A colony on a sloping
wall at Wolf Island.

Angel Chiriboga

Pocillopora meandrina.
A colony from Cousins
Island, 6 m. In this
colony and those in
the photos above, note
flattened and sinuous
branches issuing from
initial point of growth.

14

Pocillopora capitata Verrill 1864. Branches tall and upright, almost cylindrical in section, becoming flattened toward the tip. This species is close to *P.
verrucosa*, but the branches have a more thorny aspect. Verrucae elongate and
spiny, and of irregular size and distribution. **Habitat & range:** Throughout
Galápagos on shallow rocky substrates. Common at Darwin and Wolf. Eastern
Pacific distribution, possibly also Indo-Pacific.

Detail of surface. Note the spiny
edges of the irregular verrucae.
x 5.6

Pocillopora capitata from Wolf. This species can be difficult to distinguish from *P. verrucosa* without examination of the skeleton.

A typical hemispherical colony of *Pocillopora capitata* from Wolf. Verrucae are elongate and spiny.

Wolf Island

Pocillopora eydouxi Milne Edwards & Haime, 1860. Stout, upright, flattened branches that are not fused laterally. Verrucae are prominent, uniform in shape, and evenly spaced. **Habitat & range:** Occurs throughout Galápagos on rocky substrates with strong currents, usually 5-10 m depths. Uncommon central archipelago, moderately common at the northern islands of Darwin and Wolf. Mexico to mainland Ecuador and Galápagos; Indo-Pacific and far eastern tropical Pacific.

Angel Chiriboga

Left, Detail of surface. x 1.1
Above, Macro photograph of living coral from Cousins Rock, 7.5 m.

Pocillopora eydouxi, Wolf Island, east side, 8 m.

Pocillopora eydouxi,
Gordon Rocks, 8 m.
Note prominence of
verrucae.

Pocillopora eydouxi,
Guy Fawkes, 6 m.

18

Luis Vinueza

Wolf, 10 m

Pocillopora effusus Veron, 2000. Large colonies composed of prostrate flattened branches that fuse. Verrucae vary greatly in size from one colony to another. Green to brown in color. **Habitat & range:** Shallow reef environments and rocky foreshores exposed to strong wave action. Darwin and Wolf Islands in Galápagos. Rare elsewhere. An eastern Pacific species, common at Clipperton Atoll and rocky foreshores of southwestern Mexico.

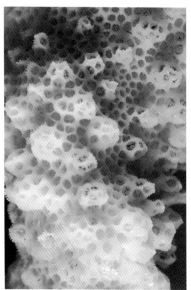

Left, Surface detail with verrucae. x 3
Above, Portion of colony showing expanded branch ends. x 1.1

Diver over a large *Pocillopora effusus* colony, Wolf 11 m.

Pocillopora effusus, Wolf anchorage, 8 m. Note the low, prostrate growth form and flattened branches, important recognition characteristics for this species.

Pocillopora effusus, Wolf east, 8 m

Pocillopora inflata Glynn, 1999. Colonies form heavy hemispherical mounds, with short, irregular, compact branches. Branches swollen at tips, a quick recognition character for the species. Verrucae acute, sparse or absent. Columellae well-developed in corallites at mid and lower branch levels. Colonies often turquoise green in color. *Pocillopora eydouxi* is similar to *P. inflata*, but the branch ends of *P. eydouxi* are never inflated, the verrucae are more uniform in size and evenly distributed over the surface, and the columellae are pinnacle-like and simple, rather than fascicular (composed of fiber bundles) as in *P. inflata*. **Habitat & range:** Shallow water, rocky areas of the central archipelago of Galápagos. Uncommon. Panamic province including Galápagos.

Left, Colony from Wolf.
Above, Surface detail. x 0.5

Pocillopora inflata colonies from Wolf. Colonies tend to be hemispherical mounds with swollen branch tips. Note also sparseness of verrucae.

22

Paul Humann

Wolf Island

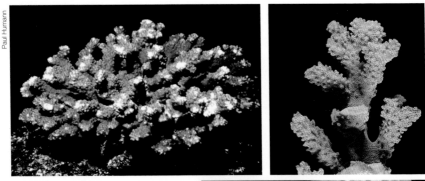

Pocillopora ligulata Dana, 1846. Colonies compact with irregular radiating branches with flattened ends and truncated tips. Similar to *P. capitata*, but differing in having verrucae widely spaced and irregular, whereas those of *P. capitata* are slightly more uniform. **Habitat & range:** Shallow water exposed to strong wave action. Presence in Galápagos requires confirmation. Indo-Pacific, Hawaii, and eastern Pacific.

Paul Humann

Above, A small colony from Wolf Island.
Above right, Skeleton of a colony, showing branching. x 0.45
Lower right, End of branch showing the irregular size and spacing of verrucae. x 2.6

Pocillopora woodjonesi Vaughan, 1918. This coral, recently recognized as present in Galápagos, is characterized by flattened branches that spread laterally over the substrate. Growth form is very similar to *Pocillopora eydouxi*. Veron (2000) cautions that *P. woodjonesi* and *P. eydouxi* are very difficult to distinguish unless the two species occur together. The branches of *P. eydouxi* are larger and more splayed. **Habitat & range:** Presence confirmed on a reef between Darwin Island and Darwin's Arch. It occurs on shallow slopes exposed to strong wave action. Throughout the Indo-Pacific and in the eastern Pacific in the Revillagigedo Islands, Mexico.

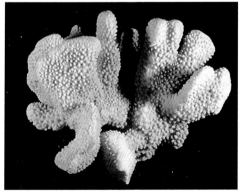

Above, Skeleton of *P. woodjonesi* from the Caroline Islands. x 0.35.
Right, Detail of surface. x 1.2

24

Phylum Cnidaria, Class Anthozoa, Subclass Hexacorallia,
Order Scleractinia, Suborder Astrocoeniina, Family Astrocoeniidae

Madracis sp. cf. *M. pharensis* (Heller, 1868). Compact colony of slightly polygonal corallites with closely opposed walls. Ten primary septa fused to the columella. This colony of colorless polyps was living on the underside of a rock and almost certainly is azooxanthellate. Two previous Galápagos records of the species *M. pharensis* are from Floreana at 30 m and Gardner Bay, Española at 64 m (Wells, 1983), both at depths that confirm the azooxanthellate nature of this coral in Galápagos, although the species *M. pharensis* is zooxanthellate on the American east coast (Veron, 2000). **Habitat & range:** On rock substrate at reported depths to 343 m. Range beyond Galápagos, Gorgona Island, Colombia (only other eastern Pacific record), eastern Atlantic and Caribbean. According

to Veron (2000) the species *M. pharensis* also occurs in the Mediterranean and northwest African coast.

Scanning electron micrograph of a five-corallite colony. x 17

Characteristics of the nine species of *Pocillopora* occurring in Galápagos

Species	Branches	Colonies	Verrucae	Range
P. damicornis	Thin	Hemispherical mound	No clear distinction between verrucae and branches	Throughout archipelago
P. verrucosa	Upright, variable	Spreading	Prominent, irregular in size	Throughout archipelago
P. ligulata	Upright, irregular, flattened toward tips	Hemispherical mound	Widely spaced and irregular	Rare, presence uncertain
P. meandrina	Flattened	Small, upright	Medium size, uniform spacing	Throughout archipelago
P. capitata	Tall, upright, nearly cylindrical in section, flattened toward tips	Hemispherical mound	Elongate and spiny, irregular in size and distribution	Throughout archipelago
P. woodjonesi	Flattened branches that fuse	Spreading	Medium size, uniform, even spacing	Darwin
P. eydouxi	Stout and flattened	Usually spreading	Prominent, elongate, uniform in size and spacing	Throughout archipelago
P. effuses	Flattened branches that fuse	Prostrate and spreading	Medium size, variable in size and spacing	Darwin and Wolf
P. inflata	Short, irregular, swollen at tips	Hemispherical mound	Acute, sparse or absent	Central archipelago

Phylum Cnidaria, Class Anthozoa, Subclass Hexacorallia,
Order Scleractinia, Suborder Fungiina, Family Agariciidae

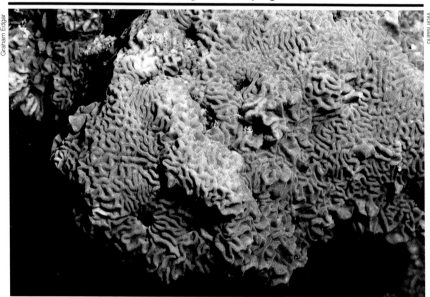

Pavona varians Verrill, 1864. Colonies laminar or encrusting. Corallites
typically are aligned in irregular valleys, but may be irregularly distributed on
flat surfaces. This is a highly variable species. Color yellow, green, or brown.
Habitat & range: Rocky substrates, widespread in Galápagos, but uncommon.
Widespread Indo-Pacific, Hawaii, eastern Pacific.

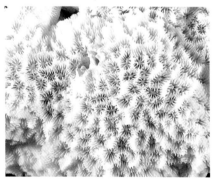

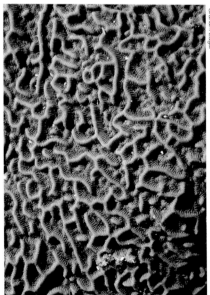

Above, Corallite structure. x 1.8
Right, Surface detail of living colony from
Beagle Rocks. Mouth and tissue color are the
same in this species.

Pavona varians.
A submassive
growth form, Wolf.

Paul Humann

Pavona varians.
Semi-laminate
growth form. As
with most herma-
typic corals, color is
variable, contrib-
uted by yellow,
green, and brown
zooxanthellae,
by red and green
filamentous algae,
and by pigments
within the coral's
own tissue.

Graham Edgar

Pavona varians.
Encrusting, semi-
laminate growth
form, Wolf.

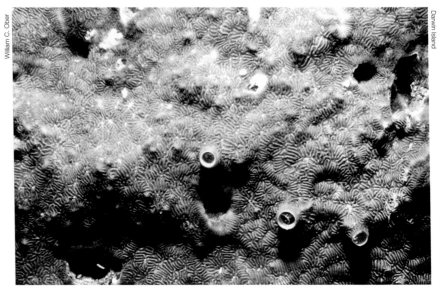

Pavona chiriquiensis Glynn, Maté, & Stemann, 2001. Colonies form flat, thin, encrusting growths, usually on basaltic substrates. Polyps and coenosarc light to dark brown. Oral disc and mouth color usually white, but may assume darker colors in deeper water with low light levels. This species is distinguished from *Pavona varians* by having short collines (skeletal ridges separating corallites) and by its encrusting growth, whereas *P. varians* has longer, more sinuous collines and often grows vertically up to 30 cm. **Habitat & range:** Prefers protected basaltic outcrops at depths of 3-20 m. Widely distributed in the eastern Pacific from Costa Rica to Ecuador and the Galápagos Islands, where it is present but uncommon at many sites in the central archipelago; moderately common at Darwin and Wolf Islands.

Left, Surface detail of corallites. x 1.8
Above, Surface detail showing short collines separating corallites.

Sylvia Earle

Wolf Island anchorage

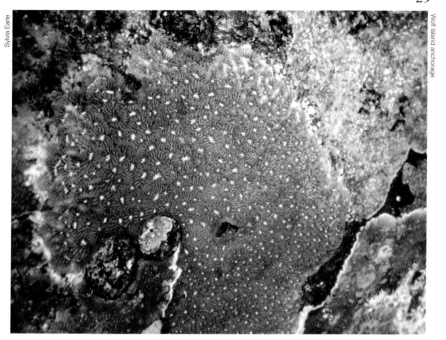

Pavona chiriquiensis with white corallite mouths, characteristic of colonies in shallow water.

Graham Edgar

Marchena

Pavona chiriquiensis from Marchena. The white corallite mouths of this deeper colony are visible but less prominent than in the shallower colony above.

Juan-Carlos Moncayo

Pavona clavus (Dana, 1846). Colonies columnar and club-shaped. Columns divide but do not fuse. Colonies can be massive. Corallites well-defined, having thick walls with septa running as fine lines from one calice to another. The coral is smooth to the touch. **Habitat & range:** In habitats exposed to current. Common throughout central archipelago and at Darwin and Wolf. Galápagos, Gulf of California, Pearl Islands, La Plata Island, Ecuador, Red Sea throughout tropical Indo-Pacific.

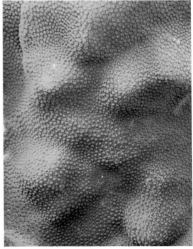

Left, Surface detail of skeleton. x 1.9
Right, Surface detail of living coral. x 0.5

Pavona clavus. Small colonies from Wolf Island developing into columnar growth.

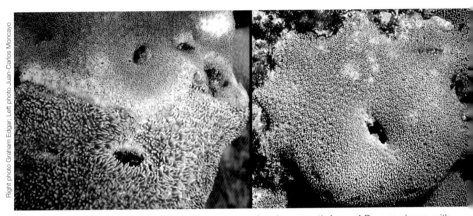

Competition for space. Encounters between *Pavona clavus* (above) and *Pavona gigantea* (below) are relatively common. Wolf Island.

Laminar growth form of *Pavona clavus,* with tentacles partly extended.

Pavona clavus. Laminar growth form.

Beagle Rocks

Pavona gigantea Verrill, 1869. Massive and encrusting species with much variation in growth form. Corallites with thick walls and well-defined columella. Polyps extended during the day giving the coral a furry appearance. It is distinguished from *Pavona clavus* by the greater distance between corallite centers, rarely less than 3 mm, and by proportionately fewer septa, generally 16 or less (Wells, 1983). **Habitat & range:** Common in shallow protected rocky substrates to depths of 25 m. Eastern Pacific Panamic province.

Juan-Carlos Moncayo

Left, Surface detail of coral skeleton. x 3.2
Above, Living coral.

 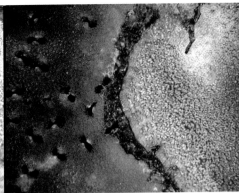

Competition for space, laminar growth. *Left*, The upper *Pavona gigantea* colony abuts the lower *Gardineroseris* colony at a "standoff zone." *Right*, A "standoff" area separates a *Porites lobata* colony on the left from a *Pavona gigantea* colony on the right. Note characteristic keyhole-shaped cavities from boring molluscs (*Lithophaga* sp.) in the *Porites* colony.

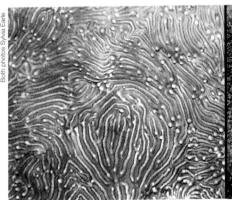

An unusual surface detail of *Pavona gigantea* showing confused septo-costae.

Pavona gigantea frequently bears the burrows created by a small hapalocarcinid crab, *Opecarinus crescentus*.

A submassive colony of *Pavona gigantea* with polyps extended, from Wolf Island.

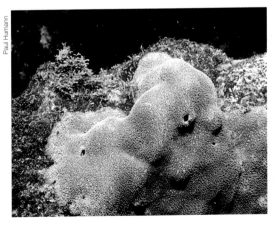

Paul Humann

Pavona maldivensis (Gardiner, 1905). Thin encrusting or columnar colonies, often mixtures of these growth forms. Thin plates grow in protected habitats, columnar and branched forms occur where wave action is strong. Corallites star-shaped, may be raised in conical form, and are of irregular size. Color grey to lavender, sometimes purple. **Habitat & range:** Relatively uncommon. Shallow rocky substrates in the central archipelago and the northern islands of Wolf and Darwin, especially areas of strong wave action. Often on vertical surfaces or entrances to crevices and caves. Mexico to Ecuador and Galápagos Islands in the eastern Pacific. Wide Indo-Pacific distribution.

Left, Detail of coral skeleton. x 2.6
Above, Living coral, showing characteristic raised corallites.

Angel Chiriboga, Darwin Island, 14 m

The encrusting plate form of this *Pavona maldivensis* colony from a vertical rock face at the mouth of a cave is a characteristic growth form in protected habitats. The columnar growth form of the colony on the facing page, upper photograph, occurs where wave action is strong.

Paul Humann

Pavona maldivensis. Thin laminar form of a colony on a vertical rock face.

Paul Humann

Wolf Island, 1992

Gardineroseris planulata (Dana, 1846). Colonies usually columnar and massive but may be encrusting. Often with laminar margins. Pale to dark brown in color. Polyps withdrawn during day. Calices polygonal, irregular, dia. 2-5 mm, closely packed. Calices larger and with higher walls than those of *Pavona* species. **Habitat & range:** Rocky substrates, often on walls or under overhangs. Galápagos populations of this species suffered greater decline from the 1982-83 and 1997-98 El Niño events than continental populations. The few isolated patches that survived the El Niño events are slowly reestablishing at Darwin and Wolf Islands as well as everywhere in the Panamic Province. Throughout tropical Indo-Pacific, Hawaii, eastern Pacific.

 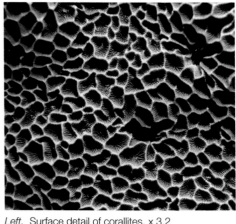

Left, Surface detail of corallites. x 3.2
Above, Living coral.

Submassive colony of *Gardineroseris planulata* at Wolf, adjacent a *Pavona gigantea* colony at right.

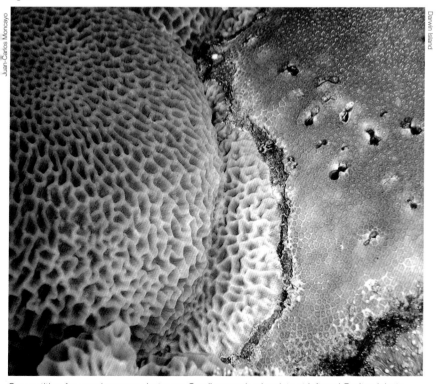

Competition for growing space between *Gardineroseris planulata* at left and *Porites lobata* at right with "standoff" area between the two species.

38

Wolf Island, 23 m

Leptoseris scabra Vaughan, 1907. Recently discovered in Galápagos, this species forms thin, encrusting colonies with well-separated, protuberant corallites inclined toward the perimeter of the colony. The surface may be highly contorted with lamina bearing corallites on one face only. Corallites are highly irregular in spacing. Septo-costa form radiating striations. **Habitat & range:** Colonies encrusting or forming thin plates on sloping or vertical walls. Often in deeper water under overhangs or in caves with reduced illumination. Restricted to the northern islands of Wolf and Darwin, where it is uncommon. Throughout the Indo-Pacific and in the eastern tropical Pacific in Galápagos, Cocos, and Clipperton Islands only. It is not found on the mainland coast.

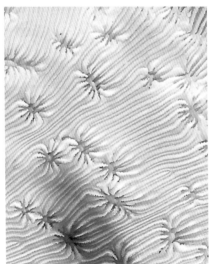 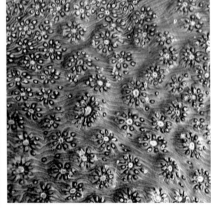

Left, Surface detail, showing interconnecting septo-costae. x 4.0
Above, Living coral. Note irregular spacing of corallites.

Leptoseris scabra from Darwin Island.

Leptoseris scabra. The raised corallites of this colony could cause confusion with *Pavona maldivensis* (see p. 34). However *P. maldivensis* lacks the radiating striations of the septa-costa that interconnect the corallites of *Leptoseris*, readily visible in this photograph.

40

Phylum Cnidaria, Class Anthozoa, Subclass Hexacorallia,
Order Scleractinia, Suborder Fungiina, Family Fungiidae

Cycloseris curvata (Hoeksema, 1989). (Center, surrounded by *Diaseris distorta*). Free-living fungiid, attached as juvenile, but living free on rubble as adult. Single corallum and polyp (retracted during day), dome shaped, slightly oval in outline, with central mouth, less than 10 cm diameter. **Habitat & range:** Coarse sand substrate east of Devil's Crown (Onslow Island) and recent sightings at Gardner Bay, Española and east side of Española. This species and *Diaseris distorta* at Devil's Crown constitute the only known remaining fungiid coral communities in Galápagos. Central Indo-Pacific, east Africa, eastern Pacific from Gulf of California to mainland Ecuador and Galápagos.

Left, Detail of skeleton. x 0.9
Above, Individual with tentacles extended. Color ranges from brown to green.

Devil's Crown, Floreana

Diaseris distorta (Michelin, 1843). Free-living fungiid composed of several fan-shaped segments, mouth located where segments diverge; polyps often inflated with water to several times volume of skeleton. Similar to *Cycloseris curvata*, but column wedge-shaped or split into wedge-shaped segments. Fragile and breaks much more easily than sturdy *C. curvata*. *Diaseris* is a mobile species, capable of excavating hollows in sand and righting itself if turned over. **Habitat & range:** Soft substrates where exposed to current. Large numbers carpet the bottom at 12-15 m in an area of strong current at Devil's Crown (Onslow Island). Throughout Indo-Pacific, eastern Africa, Hawaii, eastern Pacific from Gulf of California to mainland Ecuador and Galápagos.

Paul Humann

Left, Detail of skeleton. x 2.3
Above, Polyp with tentacles extended. Polyp color is mottled green and brown.

Phylum Cnidaria, Class Anthozoa, Subclass Hexacorallia,
Order Scleractinia, Suborder Fungiina, Family Poritidae

Wolf Island

Porites lobata Dana, 1846. Large colonies, usually helmet-shaped or hemi-spherical, surface usually smooth. Common in depths of low water to 15 m, where the colonies may reach 2 m wide and 1 m high. Polyps usually retracted during the day. Calices 1 to 1.5 mm in diameter with characteristic honeycomb appearance. This species is more resistant to bleaching from El Niño than other hermatypic corals and has recovered more rapidly than other corals following the 1982-83 and 1997-98 El Niño events. However, it was heavily impacted by cold shock from the La Niña conditions of 2007, which caused extensive bleaching (S. Banks, pers. comm.). **Habitat & range:** Shallow, stable substrates at depths to at least 30 m. Relatively common throughout Galápagos; large colonies at Darwin and Wolf. Tropical Indo-Pacific, eastern Africa, Hawaii, eastern Pacific from Gulf of California to mainland Ecuador and Galápagos.

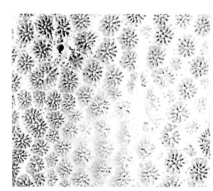

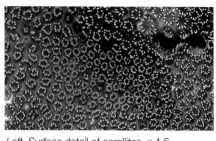

Juan-Carlos Moncayo, Pta. Espejo, Marchena

Left, Surface detail of corallites. x 4.6
Above, Surface of living coral. Polyps usually are extended only at night.

Lobular growth form of *Porites lobata* at Wolf, 9.5 m.

Luis Vinueza

Porites lobata. Edge of large colony at Sombrero Chino, 15 m.

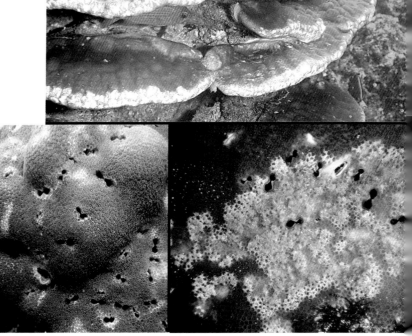

Parasitism and coral disease in *Porites lobata*. At left, a *Porites* colony bears "keyholes" produced by boring bivalves (*Lithophaga* sp.) and white bite marks from foraging parrotfish or puffers. At right, pink swellings from the disease Porites trematodiasis are caused by the encysted larval stage of a digenetic trematode. Fish complete the worm's life cycle by feeding on the pink, infected polyps. The adult worm, living in the gills of fish, releases eggs over the coral community, thus assuring reinfection of coral polyps (Vera & Banks, 2008).

NON-REEF-BUILDING (AHERMATYPIC) CORALS

Phylum Cnidaria, Class Anthozoa, Subclass Hexacorallia,
Order Scleractinia, Suborder Dendrophylliina, Family Dendrophylliidae

Paul Humann

Cladopsammia eguchii (Wells, 1982) (Syn.: *Balanophyllia eguchii*). Solitary or quasi-colonial azooxanthellate species, forming low clusters of compressed, often elongate, corallites. Columella deep, crinkled, attached to lowest inner edges of septa. Color vermillion to pinkish vermillion. **Habitat & range:** Common on walls and under rock ledges throughout the central Galápagos archipelago, from 1 to 27 m. Elsewhere, throughout the Pacific from the Gulf of Panama to Japan and Australia to 85 m.

Above, Detail of *Cladopsammia eguchii* corallite. x 9
Right, Holotype of *Cladopsammia eguchii,* USNM
46966, from Marchena, 6 m, J.W. Wells, collected
20 Feb 1975, x 0.7

Gordon Rocks, 24 m

Pinzon, 18 m

Both photos by Angel Chiriboga

Top, Cladopsammia eguchii. Polyps of this solitary coral grouped under a rock. This species succeeds in both lighted and dark environments, unlike the hermatypic corals that thrive only in sunlit environments.
Lower left and right, Cladopsammia eguchii. Close-up views of polyps with tentacles extended.

48

Cladopsammia gracilis (Milne Edwards & Haime, 1848) (Syn.: *Dendrophyllia gracilis*). Small colonies of tall, cylindrical corallites up to 36 mm tall. Calices are cylindrical to irregularly elliptical in shape and bearing tiny, triangular teeth on thin outer edges. Color of living coral is vermillion. **Habitat & range:** Common in cryptic locations, under rocks or ledges throughout central archipelago, recorded from 1 to 22 m. Elsewhere common throughout the Indowest Pacific and off the eastern coast of Australia, to 95 m.

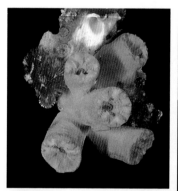

Left, Group of corallites from a colony photographed alive and as skeletons after removal of living tissue to reveal septa. Beagle Rocks, 6 m.
Above, Corallite, showing arrangement of septa. The central cavity is very deep with a central, spongy columella. x 3.5

Paul Humann

Gordon Rocks, 30 m

Rhizopsammia wellingtoni Wells, 1982. Small colonies of closely-packed corallites, more or less continually united. Calices 5 x 5 mm to 6 x 9 mm diameter. Color of polyps deep purple-black. **Habitat & Range:** Endemic to Galápagos, where it was collected from Tagus Cove, Gardner Island and Devil's Crown off Floreana, Cousins, and Gordon Rocks in sheltered locations, to 43 m. Although most colonies were destroyed by the 1982-83 El Niño event, colonies at Cousins and Gordon Rocks survived the event. Listed as Critically Endangered in the IUCN Red List (The International Union for the Conservation of Nature and Natural Resources, 2007).

Left, Holotype of *Rhizopsammia wellingtoni,* USNM 46969, J.W. Wells, Tagus Cove, 25 m, 15 June 1975. x 3.3
Right, Detail of corallite. The central, spongy columella lies deep within the calice. Corallites are united by, and budded from, the coenosteum, which spreads across the substrate. x 7

Pta. Espejo, Marchena, 8 m

Rhizopsammia verrilli Van der Horst, 1922. Spreading colonies of cylindrical corallites, which are united by stolons from which daughter polyps are budded. The presence of stolons—horizontal, cylindrical, tubular branches that pass between the bases of individual corallites—is the principal distinguishing characteristic of the species. The corallites of *Rhizopsammia verrilli* differ from those of *R. wellingtoni* in being taller (up to 1.5 cm tall), more separated with corallites joined by stolons, and with yellow-orange polyps (corallites of *R. wellingtoni* are clumped together and have purple-black polyps). **Habitat & range:** Subtidal on rocky surfaces throughout the archipelago, but uncommon (known from Wolf, Santiago, Marchena, and Floreana). Elsewhere from Cocos Island and Timor, Indonesia, to 278 m.

Left, Portion of Rhizopsammia verrilli colony. x 0.75
Right, Detail of corallite. x 2.3

Tubastraea coccinea Lesson, 1829. Colonial azooxanthellate with circular, closely-spaced corallites, many arising from a common base. Colonies brightly colored, up to 12 cm in diameter. Adult corallites 10-13 mm in diameter and up to 12 mm tall. Columella large and spongy. **Habitat & range:** Rocky walls, underhangs, avoids areas of dense coral growth. Common throughout central Galápagos archipelago and the northern islands of Wolf and Darwin, to 54 m, the ahermatype most frequently seen by casual divers. Elsewhere circumtropical to 110 m.

Left, Colonies from Champion Island photographed alive and after removal of living tissue.
Above, Detail of corallite. Corallites are the largest of Galápagos ahermatypic corals.
x 2.8

52

Caleta Negra, Isabela, 12 m

Tubastraea tagusensis Wells, 1982. Corallites cylindrical, averaging 8 mm in diameter (smaller than corallites of *T. coccinea*), closely spaced in a more or less spherical colony. This species also is distinguished from *T. coccinea* by the regular alternation of two groups of septa: a single septum lies between the twelve large, equal first and second cycle septa. Columella variable in size. **Habitat & range:** Uncommon but large colonies exist at Tagus Cove and Caleta Negra in areas of upwelling. Also recorded from Cousins Rock and Daphne Minor to 43 m. Once considered endemic to Galápagos, it was recently discovered on the coast of Brazil where it was probably introduced by ship traffic.

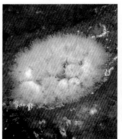

Left, Colony, showing detail of closely-spaced corallites. Colonies tend to be globular in shape.
x1.2
Above, Colony with polyps extended at left, and with polyps withdrawn at right.

Tubastraea floreana Wells, 1982. This cup coral is distinguished by small, closely-spaced, cylindrical corallites, 4-6 mm in diameter, calices 4-5 mm deep with weakly developed columella. Polyps bright pink. Corallites are smallest of *Tubastraea* species in Galápagos. Septal arrangement similar to *T. tagusensis* but corallites are much smaller. **Habitat & range:** Endemic to Galápagos on ceil ings of caves, rock overhangs. Uncommon. Playa Prieta, west side of Floreana; Gardner Island, Floreana; Caleta Iguana on rock overhang; Gardner Island, Española, in cave; Pinzón, Cousins, to 40 m. Listed as Critically Endangered in the IUCN Red List (IUCN 2007).

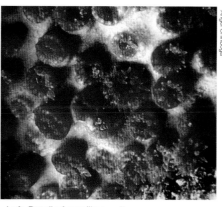

Left, Detail of corallites. x 5
Above, Colony from Gardner at Floreana
x 2

54

Tubastraea faulkneri Wells, 1982. Corallites of this uncommon coral are 8-10 mm in diameter. Because the coenosteum between calices is swollen, the calices appear to be sunken into the coenosteum surface. The first cycle of septa are prominent and exert; septa in the remaining cycles are scarcely exert or do not rise to the rim. The columella is deep and spongy. **Habitat & range:** On ceilings of caves and rock overhangs. Uncommon. Specimens collected for this study came from Gardner Island, Floreana (17 m), north Baltra (6 m), Playa Negra, west Isabela (7 m), and Sante Fe (6 m). The only specimen previously reported from Galápagos, at Tagus Cove, is lost (Cairns, 1991). Elsewhere this species has been reported from the western Pacific.

Upper left, Living colony with retracted polyps. x 1.2
Lower left, Detail of corallites after removal of living tissue. x 1.2
Above, Tubastraea faulkneri, Holotype USNM 47145, collected by D. Faulkner, July 1975, Caroline Islands. x1.5

Phylum Cnidaria, Class Anthozoa, Subclass Hexacorallia,
Order Scleractinia, Suborder Dendrophylliina, Family Rhizangiidae

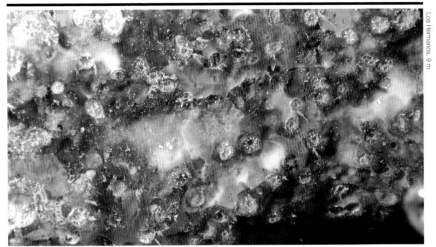

Los Hermanos, 9 m

Culicia stellata Dana, 1846. Corallites short, cylindrical but irregular, 3.5
to 5 mm in diameter. Outer edge of calice is encircled by a thin, delicate rim, a
characteristic that distinguishes this species from the two species of *Astrangia*.
The first cycle of septa protrude above the rim like teeth that, in living coral with
retracted polyps, appear white against red background tissue. The coenosteum
is white; polyps red. **Habitat & range:** Predominantly intertidal and in shallow
water to 27 m. On walls of caves and undersides of rocks, distributed throughout
the archipelago, including the northern islands of Wolf and Darwin. Elsewhere
in the Cocos Island and central and south Pacific.

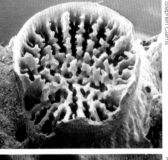

Stephen Cairns

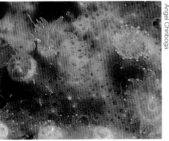

Angel Chiriboga

Above, Two corallites. Note rim encircling each
calice. x 10
Upper right, Scanning electron micrograph of single
corallite. x 13
Lower right, Portion of living colony with extended
polyps from Wolf Island.

56

Champion, 9 m

Astrangia browni Palmer, 1928. The corallites project only slightly—less than 1.0 mm—above the common coenosteum. The calices are circular to irregular, with diameters of 3.0-5.1 mm. The fossa (central cavity) is shallow with a spongy columella. The primary septa bear coarse teeth that project slightly above the rim of the calice. The columella is well developed. The coenosteum is white; polyps peach. **Habitat & range:** Cryptic, beneath rocks. Uncommon. Only two records from Galápagos: Champion Island, 2003 (this specimen) and Caleta Iguana, Isabela, 1983 (S.D. Cairns, 1991). Elsewhere, reported from Mexico.

Left, Group of corallites photographed alive (above) and after removal of living tissue (below).
Above, Detail of a corallite. The fossa is shallow and the columella poorly defined.
x 9

Astrangia equatorialis Durham & Barnard, 1952. Corallites are closely spaced and united by a common, thick coenosteum. Individual corallites project as much as 2.5 mm above the coenosteum. This species is distinguished from *Astrangia browni* by its much deeper central cavity and smaller calices (up to 3.5 mm diameter) with weakly developed columella. **Habitat & range:** Cryptic and semi-cryptic habitats on rock substrate. Uncommon. Endemic to Galápagos, throughout the archipelago including the northern islands of Darwin and Wolf.

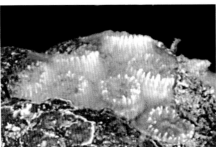

Upper left, Astrangia colonies. Colony of *A. equatorialis* at left adjacent a colony of much larger *A. browni* at right. USNM 46958, from Caleta Iguana, Isabela. x 1.8

Lower left, Lateral view of corallites. The deep central cavity (fossa) is hidden by the polyp.

Above, Detail of colony with polyps removed. Note the deep fossa and absence of a raised rim (as present in *Culicia stellata*). x 3.6

58

Oulangia bradleyi Verrill, 1866. Colonies of scattered corallites; corallites tubular, circular to elliptical, up to 16 mm diameter and 8 mm tall. This species is distinguished from other rhizangiids (*Astrangia* and *Culicia*) by its much larger size and greater number of septa. **Habitat & range:** Throughout the central and western archipelago in both exposed and cryptic habitats. Elsewhere it ranges from Mexico to mainland Ecuador. Right photograph x 4

Phylum Cnidaria, Class Anthozoa, Subclass Hexacorallia,
Order Scleractinia, Suborder Faviina, Family Oculinidae

Madrepora oculata forma *gamma* Linnaeus, 1758. A highly variable, azooxanthellate species. Corallites are solid-walled tubes linked by smooth coenosteum. The *gamma* form, one of four distinctive forms in the study material from Galápagos (Cairns, 1991) and the only one of the four found in shallow water, is an open, delicate, bushy structure occurring at depths of 15 to 24 m. **Habitat & range:** Benthic rocky substrates. A cosmopolitan species in all oceans except the Antarctic seas.

Phylum Cnidaria, Class Anthozoa, Subclass Hexacorallia,
Order Scleractinia, Suborder Caryophylliina, Family Caryophylliidae

Polycyathus isabela Wells, 1982. Colonial corallites with elliptical calice. Coenosteum and calices pink. Buds arise from thin basal encrustation, often at the base of parent corallites. The columella lies deep in the calice, consisting of 15-20 slender rods. **Habitat & range:** Endemic to Galápagos, in recesses or caves with subdued light. Uncommon. Occurrence known from Punta Albemarle (14 m, holotype, Wells, 1982); Punta Vicente Roca, 10 m; Caleta Iguana, 11 m. Listed as Vulnerable in the IUCN Red List (IUCN 2007).

Left, Detail of corallites with living tissue removed. Note the white paliform lobes, thin protrusions of all but the last cycle of septa. The presence of pali is a distinguishing character of the species. x 3.6
Above, Living colony with extended polyps from Caleta Iguana, Isabela.

60

Phyllangia consagensis (Durham & Barnard, 1952). Corallites of this azooxanthellate coral are cylindrical or irregular in shape, short, rising only 3-5 mm above coenosteum. The largest corallites are 10-11 mm in diameter. Color often a brilliant pink. **Habitat & range:** Common throughout Galápagos in caves and under ledges. Elsewhere, Gulf of California.

Upper left, Colony with tentacles with-drawn.
Lower left, Color of polyps varies among different colonies, which may be red, pink or tan.
Above, Detail of corallites. x 3

Angel Chiriboga

La Botella, Floreana, 30 m

***Caryophyllia* sp.** A solitary azooxanthellate coral attached broadly at its base to the substrate. Polyp orange with orange tentacles. The thecal edge is spiny, although this character may not be visible in the living specimen. The coral can be distinguished from *Oulangia bradleyi* by its smaller size, taller corallite and fewer septa (*O. bradleyi* has up to 96 septa in five cycles). Three species of the genus *Caryophyllia* are known from Galápagos (Cairns,1991) but because the specimen pictured was damaged during collection, it could not be identified to species. **Habitat & range:** The single specimen was on rock substrate at 30 m depth, La Botella, Floreana. Two of the three species in Galápagos also occur off Cocos Island.

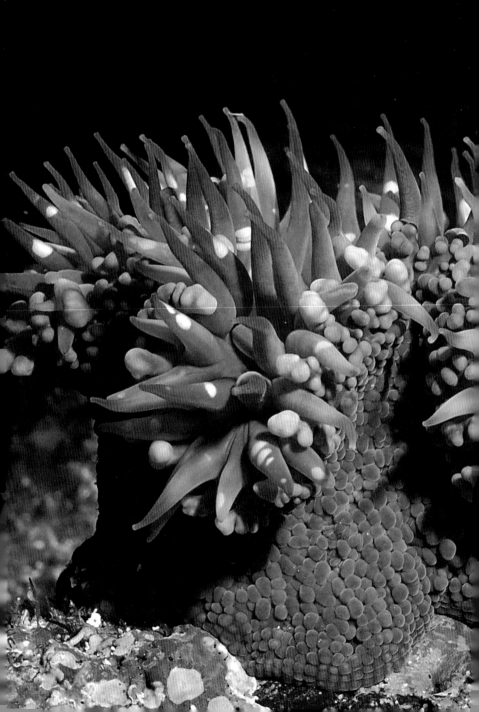

Sea anemones and zoanthids, like their close relatives the corals, belong to the subclass Hexacorallia, a group united by a body plan of six or multiples of six (hexamerous). The colorful sea anemones (order Actiniaria) are cylindrical in form with a circle of hollow tentacles crowning a stout, muscular body. They are creatures primarily of shore and shallow waters, although some species extend to all depths. In Galápagos, several species of anemones and zoanthids occur in tide pools, and some live beneath the lava rocks of the intertidal zone where they are protected from the heat of the equatorial sun. Their greatest abundance is subtidal, where they may grow in profusion on rocky slopes, attaching themselves by their pedal disc to shells, rocks, timber or any submerged strata they can find. Some anemone species burrow into the substrate.

Sea anemones are carnivorous, feeding on any animals of suitable size that they can capture on their tentacles. Because of their radial symmetry, anemones can capture food arriving from any direction. Food organisms are harpooned and immobilized by dozens of stinging capsules on tentacles, then swallowed.

Sexes are separate in many anemones whereas others are hermaphroditic. Many anemones also reproduce asexually by pulling themselves apart into halves or smaller fragments, the pieces then regenerating into small anemones.

Most widespread is the small but beautiful *Bunodosoma grandis* with its multicolored tentacles. The bright red *Phymactis clematis*, one of the largest Galápagos anemones, and one of two anemones mentioned by Charles Darwin during his visit to Galápagos in 1835, hides its brilliance by growing beneath rocks. While periodic El Niño events are damaging or lethal to many anthozoans, especially corals (see p. 2), some species use the event to gain a strong foothold in a new environment. One such anemone is *Aiptasia* sp., a clonal species that quickly spread to cover large subtidal areas in the western archipelago following the 1997-98 El Niño event.

Despite their abundance, the Galápagos anemones are not well known. Indeed, only one of several probable Galápagos endemic anemones has been described, the new species *Anthopleura mariscali* by M. Daly and D. Fautin in 2004. No

64

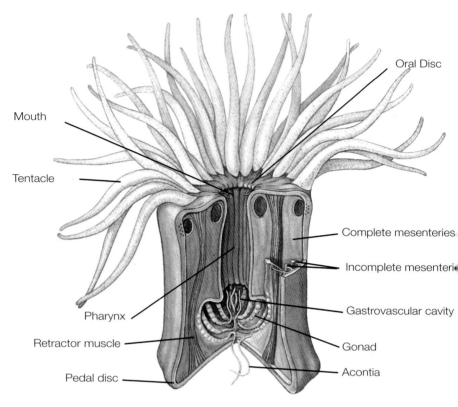

Oral Disc

Mouth

Tentacle

Complete mesenteries

Incomplete mesenteri

Pharynx

Gastrovascular cavity

Retractor muscle

Gonad

Pedal disc

Acontia

Structure of a sea anemone, cut away to show the gastrovascular cavity and mesenteries that divide the cavity into a series of radial chambers. Smaller incomplete mesenteries have mesenterial filaments bearing nematocysts and gland cells for digestion. In some anemones the filaments are extended into acontia armed with nematocysts. The acontia can be protruded through the mouth or extruded through pores in the body wall to help provide defense for the anemone.

publications of the Allan Hancock Expeditions treated the anemones. Of the approximately seventeen species of anemones that have appeared in our collections, we have been able to identify eight at least to genus with certainty; the remainder await further study.*

*The author is indebted to Drs. Daphne Fautin and Marymegan Daly for the taxonomic determinations of many of the anemones described herein.

Anthopleura mariscali Daly & Fautin, 2004. A small anemone, column cylindrical, orange near pedal disc, darkening distally, with large, adhesive verrucae. Column often covered with gravel which is shed when the animal is disturbed. Distally the verrucae abut one another in a vertical column, the terminal ones stark white, extending upward nearly as far as the marginal tentacles. Column typically 7 mm long; maximum diameter of oral disc about 15 mm. Medium-sized anemones have 48 tentacles, usually deep red but sometimes variegated white and tan. A clonal species. **Habitat & range:** High to mid intertidal on firm substrates, especially in cracks between rocks. It is sympatric with *Anthopleura nigrescens* but with a patchier distribution. Endemic to Galápagos; recorded from Santa Cruz, Santa Fe, South Plaza, and Pinzon.

Paul Humann

Bunodosoma grandis (Verrill, 1869). Column about as long as wide; entirely covered with small, non-adhesive vesicles. Surrounding the oral disc are acrorhagi, fingerlike projections bearing two to six rounded, inflated lobes, white to tan in color, contrasting with the red, orange, purple, or pink vesicles of the column. Average individuals about 20 mm in diameter with a wider, undulating oral disc (30-40 mm typically); the largest animals about 30 mm basal diameter with 60 mm wide oral disc. Tentacles very sticky to the touch, with inner tentacles longer than outer ones. Tentacle color commonly shades of brown and tan, but often beautiful hues of lavender, pink, or deep purple. A clonal species, often with several like-colored individuals packed together. **Habitat & range:** This is the most conspicuous and widespread of Galápagos sea anemones, occurring on hard substrates from mid littoral to a depth of at least 20 m, especially abundant in the shallow subtidal. Eastern Pacific distribution from Nicaragua to Peru and the Galápagos Islands, including the northern islands of Darwin and Wolf.

Bunodosoma grandis, showing diversity of color and form. At upper left, on hermit crab *Petrochirus californiensis.*

68

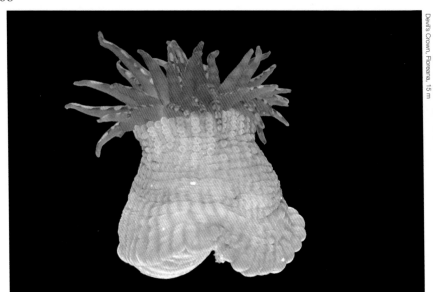

The vertical text on the right side reads "Devil's Crown, Floreana, 15 m"

Aulactinia* cf. *mexicana (Carlgren, 1951) (Syn.: *Bunodactis* cf. *mexicana*).
Small anemone, pedal disc to 1.8 cm, height 1 cm. Tentacle color variable, column pigmented from base to marginal tentacles with vertical pale orange bands. Column provided with longitudinal rows of adhesive vesicles (verrucae). Two distinct siphonoglyphs, approximately 60 rather short tentacles arranged in three or four cycles. Common to abundant in Galápagos in shallow subtidal rocky habitats. *Aulactinia* differs from *Bunodosoma* in having cup-shaped adhesive vesicles (those of *Bunodosoma* are non-adhesive) that are less dense than in *Bunodosoma*. There are no acrorhagi surrounding the oral disc. **Habitat & range:** Intertidal on and among boulders, often in areas of high wave action. Known distribution of *A. mexicana* is eastern Pacific from Gulf of California to Panama.

Anthopleura nigrescens (Verrill, 1928). A small anemone with a dark cylindrical column bearing longitudinal rows of light-colored, adhesive verrucae. Tentacles pointed, grey to tan in color with reddish cast in many individuals. The animal often partly or entirely covered with sand and gravel which tends to adhere to the verrucae, even when the animal is handled. **Habitat & range:** Common throughout the intertidal zone, attached to rocks. At upper intertidal it is confined to cracks and underside of rocks; lower down it is more exposed. In places it is sympatric with *A. mariscali* but has a wider vertical range; it also adheres less securely to the substrate than *A. mariscali*. Recorded throughout the central archipelago and north to Marchena. Beyond Galápagos this species occurs in the tropics and subtropics from north-central Indian Ocean through Southeast Asia, and in the Pacific east to Hawaii and the Galápagos, and north to Japan.

70

Phymactis papillosa (Lesson, 1830) (Syn.: *Phymactis clematis*). The largest and most distinctive Galápagos anemone. Column, tentacles, and oral disc solid bright red. As with *Bunodosoma grandis*, the column is entirely covered with small vesicles. Column length and diameter about 50 mm. Tentacles short, pointed, sticky, and longitudinally fluted when contracted. Interestingly, Charles Darwin almost certainly referred to this species when he commented on a sea anemone with "Color uniform most beautiful Lake Red." **Habitat & range:** Uncommon but widely distributed throughout the central archipelago, almost always attached to the underside of low intertidal rocks. Beyond Galápagos this species ranges from Baja California to south Chile, Juan Fernandez Island, and west to Easter Island.

Actinostella cf. *bradleyi* (Verrill, 1869). This genus is distinguished by the presence of a special wide ruffled collar outside the tentacles on an expanded disc, covered with complex rows of lobed outgrowths, but not forming true free fronds. The small, slender tentacles are withdrawn during the day. The low and broad column bears numerous adhesive verrucae in rows below the collar. There are 12 or 24 or more pairs of perfect mesenteries and one, sometimes two, endodermal sphincter muscles. Color of marginal ruff white with alternating black and light tan bands. We tentatively assign this specimen to *A. bradleyi*, first described by Verrill, 1869 as *Asteractis bradleyi*. Häussermann (2003) notes that a very similar species from the Gulf of California described by McMurrich (1893)

as *Oulactis californica*, should be assigned to the genus *Actinostella*. It is not yet known if *A. californica* is identical to *A. bradleyi*. **Habitat & range:** Intertidal on sand substrate and shallow subtidal on rock. Uncommon. Panama and Gulf of California and Galápagos Islands.

Class Anthozoa, Order Actiniaria, Family Isophellidae

Telmatactis panamensis (Verrill, 1869). A small anemone, oral disc 10-15 mm diameter, column elongate and cylindrical or, more commonly, enlarged distally with a narrow base. Most of column covered throughout with a firm, mud-colored, finely wrinkled epidermis to which sand and debris adheres. Tentacles short and club-shaped, typically 48 in number with innermost ring of 12 tentacles the largest. Tentacles usually brown or tan but occasionally purple to red. Because this is a species-rich genus with considerable variation in form and coloration among Galápagos specimens, there may be more than one species present. **Habitat & range:** On solid substrates intertidal and subtidal, to at least greatest SCUBA depths (35 m). Always shielded from direct sunlight, usually on underside of rocks. Throughout archipelago including Darwin and Wolf. Eastern Pacific south to Chile.

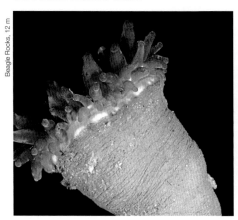

Caleta Iguana, Isabela, 12 m

Calliactis polypus (Forskål, 1775). This anemone lives on shells inhabited by hermit crabs. The base of column broadly expanded, adhering firmly to the hermit crab shell. Column broad and moderately elevated in expansion; flattened to a low cone when contracted (as is the specimen at left in the photograph above). Column nearly smooth except near base where there is a conspicuous single row of elevated, white cinclides forming a circle around the anemone. Extending proximally from this circle to the base are alternating white and dark pigmented bands. Tentacles numerous, slender, and highly contractile. **Habitat & range:** Almost always found on the shells inhabited by hermit crabs, in Galápagos usually *Dardanus sinistripes* but occasionally on *Petrochirus californiensis*. Occurs most frequently on soft substrates in bays, coves and harbors to at least 40 m. The hermit crab actively transfers one to three anemones to its shell. In this symbiotic relationship, the hermit crab benefits from camouflage, protection from predators, and coverage of weak or perforated shells; the anemone is protected in this relationship from some predators, such as polychaetes, nudibranchs, certain sea stars, and some fishes. Cosmopolitan distribution.

Paul Humann

Wolf, 15 m

The hermit crab *Dardanus sinistripes* commonly bears multiple *Calliactis* anemones on its shell.

74

Antiparactis* cf. *lineolatus (Couthouy in Dana, 1846). Small clonal anemone, almost always on branches of black coral or gorgonians. Column smooth, tentacles in two or three cycles depending on anemone size, with largest tentacles in innermost cycle. Of the two morphs collected in Galápagos, the column of the "leopard" form is marked with irregular chocolate-brown spots; this form most closely follows McMurrich's (1893) description of the species. The body wall of the other morph, which we tentatively include with this species, is red or white in color, lacks "leopard" spots, but is pebbled with numerous minute white or dark pigment spots. Judging from the close association of individuals, this is a clonal species that reproduces by pedal laceration. **Habitat & range:** Subtidal on branches of black coral and gorgonians, usually, but not always, below 20 m. In Galápagos we have recorded this species at Cape Marshall (Isabela), Roca Redonda, Gordon Rocks, Bainbridge Rocks, Darwin (Darwin's Arch), Wolf, and Genovesa. McMurrich (1893, 1904) reported this species from Panama and Juan Fernández Island, Chile.

Antiparactis cf. *lineolatus* from different Galápagos sites, showing variation in pigmentation patterns.

Punta Espinosa, Fernandina, 5 m

***Aiptasia* sp.** A clonal species living in densely-packed monocultures in which they appear almost translucent. Extensible column brown and hour-glass shaped, usually only the oral disc visible. Oral disc 15-20 mm in diameter; mouth white surrounded by a brown ring. About 70 narrow, tapering tentacles occur only at margin, color yellowish-brown. **Habitat & range:** Forming continuous carpets on rock substrates to 5 m depth. This species was first reported from Espinosa Point as new to Galápagos during the 1997-98 El Niño event; however, it may have appeared earlier (Paul Humann, pers. comm.). By 2007 the species had spread throughout the west and central archipelago, often appearing at high density in areas of heavy fishing pressure (Sonnenholzner, et al. 2007).

Punta Espinosa, Fernandina, 4 m

ZOANTHIDS

78

*Z*oanthids are hexacorals, soft-bodied anthozoans that superficially resemble small sea anemones, but most are colonial and are united at their bases. Zoanthids are common in Galápagos. Most species reproduce primarily asexually to produce colorful, spreading colonies that can be quite extensive, such as the species seen in Galápagos. Although solitary zoanthids exist, all of the Galápagos zoanthids that we have observed are colonial. Some zoanthids have symbiotic zooxanthellae in their tissues, while others are primarily heterotrophs that ingest food captured on their tentacles and presumably can survive without autotrophy.

The tentacles develop in two rings at the edge of the oral disc and usually number in multiples of six. A single siphonoglyph lined with cilia aids in feeding and water exchange. The body wall is a leathery cuticle that provides some protection for these soft-bodied animals. Species of *Palythoa* and probably species of other Galápagos zoanthids contain a potent neurotoxin in their tissues, called palytoxin, which serves as a defense against predation by crabs and fish; some predaceous snails, however, are immune to the toxin.

Despite near ubiquity of zoanthids in marine seas, the group remains relatively unknown and neglected by zoological researchers. While many genera are well established, the extent of species diversity remains obscure. Recent examinations have shown a hitherto unknown level of diversity within the zoanthids, and taxonomic revision is ongoing. Zoanthids commonly seen in the shallow waters of the tropical world are currently placed in four families, two of which are primarily zooxanthellate (Zoanthidae, including the genus *Zoanthus*, and Sphenopidae, including the genus *Palythoa*) two of which are primarily azooxanthellate (Parazoanthidae, including the genera *Parazoanthus* and *Savalia*, and Epizoanthidae, containing the single genus *Epizoanthus*). The identification of zoanthids, even to genus, is never simple due to plasticity in colony size, polyp shape, color, and epizootic host, among morphological and ecological variables. Consequently, we applied both morphological (classical histology) and molecular approaches to determine the identity of Galápagos zoanthids.

The author is indebted to James D. Reimer, University of Ryukyus, Okinawa, for the molecular determinations of zoanthid genera reported here and to John S. Ryland, University of Wales Swansea, for assistance with the interpretation of histological preparations.

We have identified three genera of zoanthids in Galápagos with no azooxanthellate zoanthid identified to definitive species. Many of our samples are almost certainly undescribed new species. Zooxanthellate *Zoanthus* species grow prolifically on intertidal flats, forming extensive colorful clonal patches on rock substrate.

Palythoa, like *Zoanthus*, is zooxanthellate and found on wave-washed low intertidal habitats. Unlike *Zoanthus* and like other zoanthids, *Palythoa* species take up sand and detritus to help form their structure. As mentioned above, some *Palythoa* species contain a neurotoxin which, in fact, is one of the most potent toxins known.

With one exception, the observed subtidal zoanthids in Galápagos are azooxanthellate Parazoanthidae species. While all observed specimens are currently included in the family Parazoanthidae, recent work using DNA has suggested two groups to be clearly distinct from *Parazoanthus* sensu stricto (here designated *Parazoanthus* sp. G2). This clade generally associates with sponges, although we have seen Galápagos colonies spreading and growing directly on rock substrate. *Parazoanthus* sp. G1 was observed to be epizoic on antipatharians in Galápagos, while *Parazoanthus* sp. G3 was usually observed attached to rock substrate, not epizoic on hydrozoans as many of this clade apparently are. As mentioned above, taxonomic revision within the zoanthids is ongoing, and it is likely that both Parazoanthus spp. G1 and G3 will be redescribed in the near future as new genera or families.

With the help of John Ryland, we identified many specimens of *Epizoanthus* among the Galápagos specimens prepared for histological study. Unexpectedly however, the molecular methods applied by James Reimer failed to identify the genus *Epizoanthus* among the Galápagos specimens. The Parazoanthidae and Epizoanthidae are notoriously difficult to distinguish morphologically; separation of the two depends on whether the sphincter muscle is endodermal (Parazoanthidae) or mesogleal (Epizoanthidae). Despite their absence in our collections, it is possible that species of Epizoanthidae may occur at depths beyond those normally explored by SCUBA, or among still-unclassified zoanthid specimens.

Phylum Cnidaria, Class Anthozoa, Subclass Hexacorallia,
Order Zoanthidea, Family Zoanthidae

Zoanthus **cf.** *sansibaricus* Carlgren, 1900. Species of this intertidal
and shallow-water zooxanthellate zoanthid have wide, flattened oral discs
surrounded by a double ring of short tentacles. This is the only Galápagos
zoanthid that does not have sand and other sediment incorporated into the body
wall. The genus is distinguished by internal characteristics: a) brachycnemic
mesenteries (the fifth couple of mesenteries counting from the dorsal directives
are incomplete, i.e., they do not extend from the body wall to the pharynx) and,
b) a divided mesogleal sphincter muscle. They reproduce asexually by budding
to form spreading, stoloniferous mats, and also spawn annually similar to many
hermatypic corals. The large majority of Galápagos *Zoanthus* specimens are
either *Zoanthus sansibaricus,* Carlgren, 1900, common in the Indo-Pacific, or
a very closely related species. *Zoanthus vietnamensis* Pax & Mueller, 1957,
distinguishable by their pale pink or purple oral disks, may also be present
in small numbers on Santa Cruz (Reimer, pers. com.). **Habitat & range:**
Species of this zooxanthellate genus grow prolifically in shallow, low-littoral
areas of intertidal flats and extend into shallow, low-energy shorelines. Large
colonies are seen near the Charles Darwin Research Station, at Tortuga Bay, and
elsewhere along the southern coast of Santa Cruz. We also found a single colony
at Point Espejo on Marchena. Similar to hermatypic corals, *Zoanthus* spp. are
limited to warmer waters where their zooxanthellae can survive (i.e. minimum
temperatures >15.0° C). Green, brown and other color morphs are often found
intermixed. The type locality of this species is Zanzibar; it ranges widely over
subtroptical/tropical Indian and Pacific Oceans.

Zoanthus cf. *sansibaricus* polyps removed from colonial mass, showing that unlike all other Galápagos zoanthids, this species has no sediment embedded in body wall.

Edge of expanding colony of *Zoanthus* cf. *sansibaricus* with polyps of varying age and size.

82

Phylum Cnidaria, Class Anthozoa, Subclass Hexacorallia,
Order Zoanthidea, Family Sphenopidae

William C. Ober

Santa Cruz, intertidal

Palythoa mutuki Carlgren, 1937. This species can be distinguished from
P. tuberculosa with which it is closely related by its well-developed polyps
extending free and clear of the coenenchyme. Oral disks are often green or
brown in color. Often found intermixed with *Zoanthus* colonies, polyps are
brown or tan on their external surface, and oral disks are approximately twice as
large as *Zoanthus* (approx. 1-1.5 cm in diameter). Colonies contain a powerful
neurotoxin and should not be touched. **Habitat & range:** Similar to the other
zooxanthellate species, *P. mutuki* is found in warm-water, wave-washed infra-
littoral zones on Santa Cruz and likely other areas. Widespread through the
tropical and subtropical waters of the Indo-Pacific Ocean.

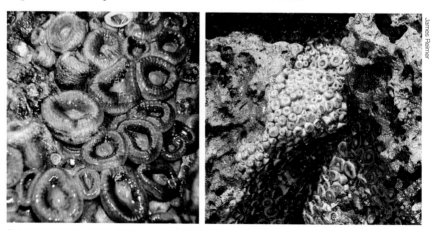

James Reimer

Palythoa mutuki colonies in tide pools at (*left*) Tortuga Bay, Santa Cruz and (*right*) a beach
near Charles Darwin Research Station.

Large colony of *Palythoa mutuki* in a tide pool at Tortuga Bay, Santa Cruz.

Palythoa tuberculosa Delage & Herouard, 1901. The polyps of this warm-water, zooxanthellate species barely extend from a well-developed coenenchyme that is impregnated with sand particles. Polyps are packed tightly together, forming rubber-like mats. Color is pale yellow or buff. Like *Zoanthus*, *P. tuberculosa* prefers wave-washed intertidal habitats, but colonies do not spread as extensively as those of *Zoanthus*. **Habitat & range:** Primarily intertidal. Polyps seem usually to be contracted during the day. Colonies should not be touched, as they contain high concentrations of a powerful neurotoxin. This species is quite common throughout the Indo-Pacific.

Palythoa tuberculosa collected from Santa Cruz, east of the Darwin Station.

84

Phylum Cnidaria, Class Anthozoa, Subclass Hexacorallia,
Order Zoanthidea, Family Parazoanthidae

Cousins Rock, 20 m, on dead black coral

***Parazoanthus* sp. G1 (antipitharian-associated)** *Parazoanthus*
sp. G1 has bright yellow, red, or white tentacles, with long red, yellow, or
cream colored polyps that extend well clear of the coenenchyme. Similar to
Parazoanthus sp. G3, this species belongs to a clade that is likely to become
a new genus in the near future. It is characterized by being epizoic on
antipatharians (black coral), and nearly all collected samples from Galápagos
were closely associated with the black coral *Antipathes galapagensis*. Examined
specimens are apparently from a single species, found at depths of 6-25 m. The

polyps may cover only a portion of
a living black coral colony, or cover
the entire colony, thus killing the
anitpatharian and suggesting this
species may be somewhat parasitic.
Habitat & range: Epizoic on
black coral. Although black coral is
found throughout the archipelago,
Parazoanthus sp. G1 colonies were
observed only at Isabela (Las Marielas
& Elizabeth Bay), Cousins Rock, and
Floreana (La Botella); it may be that
this genus has a patchy distribution in
the Galápagos.

Polyps of *Parazoanthus* sp. G1

Parazoanthus sp. G1 on black coral, showing variation in tentacle and polyp color.

Caleta Iguana, Isabela, 9 m, on rock

***Parazoanthus* sp. G2 (sponge-associated)** The genus *Parazoanthus* has been a catch-all for many unknown or undescribed species of azooxanthellate zoanthids. However, recent examinations of DNA are leading to a reorganization of the former genus *Parazoanthus*, and in the future it is likely that the genus will consist only of species that associate with, or are epizoic on, sponges (*Parazoanthus* sensu stricto). Collected *Parazoanthus* sp. G2 specimens from Galápagos are often associated with a bright orange sponge, possibly *Pseudosuberites* or *Tedania*. *Parazoanthus* sp. G2 colonies often grow in patches over the sponge, or may even cover it entirely, often extending to surrounding rock substrate. Despite being covered by the zoanthid the sponge is always alive, suggesting this association is symbiotic and not parasitic. *Parazoanthus* sp. G2 polyps have yellow, orange, or cream tentacles, and a red, yellow, or light yellow oral disk, with a light tan, light pink, or cream coenenchyme. Although polyps extend clear of the coenenchyme, when contracted the polyps are mere bumps on the surface of the coenenchyme, much more embedded than either *Parazoanthus* sp. G1 or sp. G3. Colonies may be very small, or extend to cover more than a square meter in area. It is highly likely that all collected specimens of the sponge-associated *Parazoanthus* from Galápagos belong to a single species as yet undetermined. **Habitat & range:** Associated with sponges. Similar to *Parazoanthus* sp. G1, *Parazoanthus* sp. G2 zoanthids are found on rock walls, in crevices, or at the base of rocks, and occur from depths of 5 m to the lower limit of SCUBA, and may extend deeper. *Parazoanthus* sp. G2 was seen at Isabela, Fernandina, Floreana, and Gordon Rocks and its range is likely throughout the entire archipelago. The range of this species beyond Galápagos is unknown.

Parazoanthus sp. G2 colony on red sponge (possibly *Tedania* sp.). Polyps are fully embedded in the sponge.

When polyps are fully contracted, as in this colony, polyps are reduced to rounded swellings on the coenenchyme.

Small *Parazoanthus* sp. G2 colony growing with sponge on rock.

88

Pta. Vicente Roca, wall, 9 m

***Parazoanthus* sp. G3** Most colonies of this species have red or red-brown oral disks. The outer surface of polyps are tan to dark brown, with polyps more well-developed and clear of the coenenchyme than *Parazoanthus* sp. G2 (sponge-associated), but less well-developed than *Parazoanthus* sp. G1 (antipatharian-associated). *Parazoanthus* sp. G3 belongs to a clade that is potentially an undescribed genus or family, and that associates closely with, or is epizoic on, hydrozoans. However, most specimens observed in Galápagos were living on rock substrates. To our knowledge, this is the only Pacific zoanthid species that usually is non-epizoic and bright red in color. Two specimens had white oral disks (bottom photograph, p. 89) but were morphologically and molecularly similar to red specimens. Colonies may spread over an area greater than a meter in diameter. It is likely the collected specimens from the Galápagos represent a single species. **Habitat & range:** Colonies generally present at the

Parazoanthus sp. G3 growing on dead coral. Darwin Island, 8 m depth.

base of large rocks and on rock walls in areas of high current. We found colonies at Darwin, Pinzon, Genovesa, Marchena, Isabela (Punta Vicente Roca, Caleta Iguana, Elizabeth Bay), Española and Floreana. It is probable that this species is found throughout the archipelago. It occurs from low infra-littoral to depths of over 35 m, and probably can be found at even deeper depths.

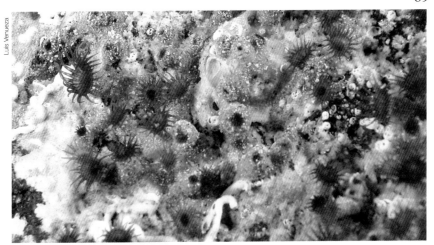

Parazoanthus sp. G3 colony from Roca Onan pinnacle, Pinzon.

Parazoanthus sp. G3 specimen shown with tentacles expanded and contracted (inset).

Parazoanthus sp. G3 with white polyps. Genovesa at entrance on east side, 9 m, on empty mollusc shells. This white form of the clade may be a different species from the much more common red forms.

Unidentified zoanthid. This small zoanthid is characterized by a heavy encrustation of coral sand and detritus particles, especially on the coenenchyme (stolons) connecting the individual polyps. The polyps are among the smallest of Galápagos zoanthids, with diameters of 1-2 mm, oral disc diameters of 1-3 mm and polyps no more than 4 mm in length. The dark oral disc bears 24-36 tentacles that are approximately twice as long as oral disc diameter. The polyps are barely connected to each other by a thin coenenchyme and are sometimes independent. Molecular examination of this unidentified zoanthid reveals it to be a novel form distinct from other zoanthids in the Parazoanthidae clade.
Habitat & range: Specimens were found on the underside of rocks or dead coral fragments resting on sandy bottom at depths of 6 to 27 m. Some specimens were growing over encrusting bryozoans; others were not associated with any particular fauna. We

have found specimens at Darwin, Wolf, Marchena, Santiago, North Seymour, San Cristóbal, Española, and Floreana Islands.

Unidentified zoanthid with polyps in contracted state.

CERIANTHIDS
BLACK CORAL
CTENOPHORES

CERIANTHIDS - TUBE ANEMONES

The cerianthids, or tube anemones, are among the most beautiful and graceful of radiates. They are large, solitary, elongate animals that live in vertical tubes in soft substrates. The tentacles are arranged in two sets, a marginal set of long thin tentacles used for prey capture and an oral set of fewer, shorter tentacles encircling the mouth used for prey manipulation and ingestion; each set may contain one or more circlets (pseudocycles). There is much color variation in the tentacles of most cerianthids. The long, tapering body is thrust into its secreted tube, composed of mucus and specialized, cast-off nematocysts, or embedded with sand grains and small pebbles. When disturbed, which may require no more than a passing shadow, the animal can retract into its tube in a flash.

Despite their appearance and common name, tube anemones are not anemones at all. Their internal anatomy is completely unlike that of true anemones. They are, in fact, more closely related to black corals than to true sea anemones.

Cerianthids feed mainly on plankton and organic debris captured on the tentacles, but may occasionally snare small fish with the stinging cells of their tentacles. Cerianthids are noted for their long lives. One individual in the Naples Aquarium lived for more than 40 years, increasing ten times in size.

Cerianthids are widely distributed in tropical and subtropical seas from shallow water to great depths. Most widely distributed in Galápagos is *Botruanthus benedeni*. An extensive population of this species shares the sandy bottom of Tagus Cove with the U-shaped tube worm *Chaetopterus* sp.

Phylum Cnidaria, Class Anthozoa, Subclass Hexacorallia,
Order Ceriantharia, Family Cerianthidae

Botruanthus benedeni (Torrey & Kleeberger, 1909) (Syn.: *Cerianthus benedeni*). This burrowing tube anemone is readily distinguished by the numerous marginal tentacles arranged in three pseudocycles. The tentacles of the outer cycle thinner and somewhat shorter than those of the middle and innermost cycles. Tentacle color is variable, ranging from near-colorless to dark purple. Surrounding the mouth are numerous delicate labial tentacles. The body is thick walled and muscular. The tube into which the anemone can instantaneously and completely retract when disturbed is 3-5 mm thick, silky and pliable in texture, and may contain scattered particles of debris. **Habitat & range:** Sandy substrates in gentle current, often in association with *Chaetopterus* sp. A large, multicolored population exists on the floor of Tagus Cove from 12 to 30 m depth. Other colonies exist in James Bay, Devil's Crown, and Rabida and are probably widely distributed in the central archipelago. Southern California to Ecuador and the Galápagos Islands.

Specimens of *Botruanthus benedini* from Tagus Cove and Devil's Crown showing color variation.

Phylum Cnidaria, Class Anthozoa, Subclass Hexacorallia,
Order Ceriantharia, Family Arachnactidae

Paul Humann

Rabida

***Arachnanthus* sp.** A small cerianthid with slightly curved marginal tentacles
arrayed at the margin of the oral disc in one or two pseudocycles. The labial
tentacles form a cone-like structure. Color of marginal tentacles variable.
Banding of tentacles is characteristic with bands at the tip and base of a tentacle
commonly darker than those in the middle of the tentacle. Labial tentacles are
brown or tan. The oral disc is white or tannish. The tube of *Arachnanthus* is thin
with the consistency of cardboard. **Habitat & range**: Buried in sandy substrates.
Less common than *Botruanthus benedeni*. Individuals have been sighted at
Rabida and Punta Vicente Roca. The genus is cosmopolitan.

J. Green, Punta Vicente Roca, Isabela, 8 m

BLACK CORAL

B lack corals, also called thorny corals, take their common
names from the black or brown color and spiny surface
of the tough, proteinaceous internal skeleton. The branch-
like colonies, ranging from a few centimeters to several meters
in length, are attached at their base to a firm substrate. Like
gorgonians, the skeletal axis is covered by a living coenenchyme
bearing small polyps, each with six non-retractable, pinnate tentacles covered with nematocysts. The polyps are dioecious but the
colony may be hermaphroditic.

The order name, Antipatharia (Gr., *anti-,* against, remedy
for + *pathos,* suffering) derives from an ancient notion, still widely
held in the tropics, that black coral has healing power. The black
skeleton is cut and formed into bracelets worn in the vain hope that
they will cure afflictions such as arthritis.

Black corals are most abundant in the deep waters of tropical and subtropical oceans. The two black coral species in Galápagos occur throughout the archipelago from 3 to at least 90 m depth,
with *Myriopathes panamensis* found in the upper part of this range,
and *Antipathes galapagensis* more common in the lower part. Both
species reportedly grow more slowly in Galápagos than in Hawaii, where growth rate is 6.1 cm/year and life spans may exceed
80 years. After death, the skeleton of black coral serves for many
years as a platform for many kinds of sessile marine life.

Once collected in Galápagos for the jewelry trade, black
coral colonies suffered consid-
erable damage that will take
many years to repair. Collecting
is no longer permitted within
the Marine Reserve.

Tentacle

Spiny skeleton

96

Antipathes galapagensis Deichmann, 1941. Tall, bushy colony of
elongate branches of varying thickness
with non-retractable polyps scattered
irregularly on the stems and branches.
Colonies reach 1 m in height. Living
colonies dull to bright yellow. **Habitat
& range:** Slanting rocky substrates and
ledges at depths of 5 m to more than 50
m. Best developed colonies occur below
20 m and in light current rather than
strong current. In Galápagos this species
is more common than *Myriopathes panamensis*. From the Gulf of California to
Ecuador and the Galápagos Islands.

Phylum Cnidaria, Class Anthozoa, Subclass Ceriantipatharia,
Order Antipatharia, Family Myriopathidae

William C. Ober

Tagus Cove, 4.5 m

Myriopathes panamensis (Verrill,
1869) (Syn.: *Antipathes panamensis*).
Branching colony in thick, multilayered
plane, complexly pinnate with pinnules
4-6 mm long, arranged bilaterally and
alternately. Color yellow-orange. The pinnules are studded with 7-9 rows of minute
spines. Non-retractable polyps are small,
white and arranged along the pinnules with
10 to 12 polyps per cm. Reproduction is
both sexual and asexual, the latter by colony fragmentation. Formerly harvested for
black coral jewelry, the collection of this
species in Galápagos is now forbidden.
Habitat & range: On vertical rocky substrates in weak current. Depth 3-50 m with
optimum development at 15 m. Density in
Galápagos is approximately one-tenth that
of *Antipathes galapagensis*. From the Gulf
of Panama and the Galápagos Islands.

CTENOPHORES

Ctenophores (pronounced teen'o-fours), also called comb jellies and sea walnuts, are a marine group composed of fewer than 100 species. The phylum name Ctenophora (Gr. *ktenos*, comb, + *phora*, pl. of bearing) derives from the eight rows of ciliated comb-like plates they use for locomotion. Light-scattering of sunlight produced by beating of the comb plates provides a beautiful changing rainbow of colors passing down the rows. However, their small size (most are less than 3 cm in diameter) and fragile, transparent bodies make them inconspicuous during daylight hours. Many ctenophores, although not *Hormiphora,* are more easily seen at night when they emit blue or green light by bioluminescence.

Cydippid ctenophores, such as species of the genus *Hormiphora,* have two very long and extensible tentacles that may be retracted into a pair of tentacle sheaths. Lacking nematocysts, the surface of tentacles bears colloblasts; these are glue cells that produce a sticky substance used in capturing and holding small animals. A prey-laden tentacle contracts to bring food near the mouth for ingestion. Despite their delicate appearance, ctenophores can be remarkably effective predators of zooplankton and larval fishes.

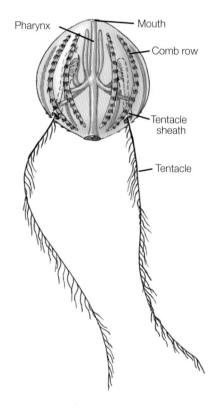

Pharynx

Mouth

Comb row

Tentacle sheath

Tentacle

Phylum Ctenophora, Class Tentaculata, Order Cydippida,
Family Pleurobrachiidae

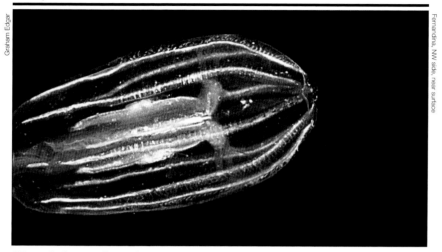

Hormiphora* cf. *palmata Chun, 1898. A spindle-shaped, transparent
ctenophore, soft and flaccid, body slightly flattened. The comb rows run 2/3
to 4/5 the distance from the aboral pole, are evenly spaced and about the same
length. Tentacle sheaths large and angle toward the aboral pole; tentacles (not
visible in this photograph) are long. *H. palmata* is widely distributed and known
from Hawaii and Mexico.

***Hormiphora* sp. A.** Body transparent and ovate, about 1 1/2 times longer
than wide and circular in cross section. Comb rows equal and evenly spaced.
Long tentacular sheaths lie close to the pharynx for much of their length, then
exit near the aboral pole. Tentacles (one partially intact in photograph) capable
of great extension. Similar in appearance to *Pleurobrachia* sp., it is described
only in an unpublished M.A. thesis (1931) under the nomen nudem *H. coeca*.

GORGONIANS
SEA PENS

Gorgonians and sea pens belong to the subclass Octocorallia, radiates that are built on a strict octomerous symmetry. The polyps have eight pinnately branched tentacles with eight unpaired, complete septae. The branching structure of a gorgonian colony (order Gorgonacea) is supported by a flexible core of a tough, horny material, called gorgonin, over which is spread a thick, gelatinous coenenchyme (seen'enkyme) containing minute skeletal elements, the spicules. Spicules of various sizes and shapes, are important in the taxonomy of gorgonians. The polyps, armed with stinging cells (nematocysts), are quite effective in capturing small planktonic organisms in large numbers. They are retractile, capable of withdrawing into the coenenchymal mass, or into an elevated mound on the coenenchyme, the calice.

Gorgonians harbor skeletal elements in their tissues, called sclerites, which provide support against wave action and structural defence against predators. Differences in sclerite shape, ornamentation, and size are characteristic of the species and thus serve as important diagnostic characters.

As with the hermatypic corals, the brown and green colors of many gorgonians are due to the presence of zooxanthellae in their tissues. Others derive their brilliant red, yellow, orange and lavender colors from pigments in their sclerites. The graceful, plantlike form of gorgonians, enhanced by these vivid colors, contribute much to the allure of submarine gardens. All gorgonians are found below the low tide mark in areas of mild current. Most are so sensitive to strong light that the polyps are found fully extended only at night or on sunless days. Living in moving water as they are, gorgonians make an inviting support for a host of invertebrate larvae of sponges, brachiopods, shrimp, snails, and hydroids. In Galápagos, the branches of sea fans of the genus *Pacifigorgia* are often almost completely obscured by thousands of epizoic brittle stars (*Ophiothela mirabilis*) that use gorgonians as convenient feeding stations for trapping plankton passing by in the current—and competing with their gorgonian hosts, who also feed on captured zooplankton. Gorgonians also feed by digesting their

The author is indebted to Odalisca Breedy, who accompanied the author on one research trip, for taxonomic determination of gorgonians described in this section. He is indebted to Gary Williams for identification of the sea pens in this section.

zooxanthellae and absorbing dissolved organic matter from sea water.

The sea pens of the order Pennatulacea have evolved a highly specialized type of symmetrical colony developed around a long, primary polyp; the lower end may be dilated to serve as an anchor in mud or sand. Sea pens are the only octocorallians adapted to live in soft substrates. Before the 1982-83 El Niño event, at least three pen species could be found on soft substrates. Today, finding a sea pen anywhere in Galápagos is a rare occurrence.

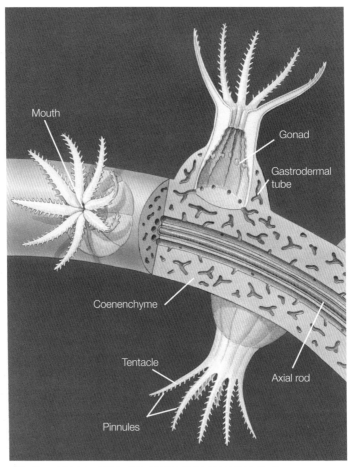

Structure of a gorgonian, showing the eight pennate tentacles, coenenchyme, and the horny axial support.

Class Anthozoa, Subclass Octocorallia, Order Gorgonacea,
Family Plexauridae

Muricea **cf.** *fruticosa* Verrill, 1869. This stocky, small gorgonian has
branches that subdivide repeatedly into stiff branchlets, studded throughout with
prominent calices. Extending from the calices are white polyps. The branches
and calices are reddish-brown on the outermost branches but fade to white on
the inner, older branches. Colonies in Galápagos seldom exceed 15 cm in diameter. **Habitat & range:** Secured to rock faces mainly where exposed to moderate current. Unevenly common throughout the central archipelago and Darwin
Island. Distributed from Panama to Ecuador and the Galápagos Islands.

Left, Enlarged view
of branches with
expanded white polyps
with red mouths.

104

Pta. Vicente Roca, 7.5 m

***Muricea* sp. 1** The trunk of this large sea fan separates at its base into two or more main branches which give off secondary branches, these subdividing repeatedly in large specimens. The branches and branchlets, which tend to grow in one plane, are thick and closely applied to each other, forming an almost solid fan in some specimens. Color ranges from brown to deep red. This is one of the larger gorgonians present in Galápagos, reaching colony diameter of 0.5 m or more. We believe this is a new species (O. Breedy, pers. comm) that is similar to, but different from, *Muricea appressa* (Verrill, 1869). **Habitat & range:** Widespread throughout the central archipelago. Always strongly attached to solid substrates in areas of moderate current. Known distribution Mexico, Panama, Ecuador, and Peru.

Angel Chiriboga

Caleta Iguana, Isabela, 21 m

Closeup views of branches of *Muricea* sp. 1 showing expanded polyps.

The polyps of *Muricea* sp. 1 instantly retract when touched. This specimen is shown before and after it was gently brushed with the author's hand.

Some examples of variation in color and growth form in *Muricea* sp. 1. From upper left clockwise, Tortuga Island; Cape Douglas, Fernandina; Albemarle Point, Isabela; La Botella, Floreana.

106

Point Pitt, San Cristobal

***Muricea* sp. 2** This gorgonian resembles *Muricea fruticosa* in size and mode
of branching, with moderately stout branches that divide and subdivide at inter-
vals. The calices are prominent and armed with sharp spicules. The polyps are
transparent and inconspicuous. This gorgonian appears in several chromotypes.
It bears resemblance to Verrill's 1869 description of *Muricea formosa,* from a
single all-white specimen collected at Zorritos, Peru. However, the identity of
this multi-pigmented Galápagos gorgonian must await further study. **Habitat
& range:** Subtidal on rock surfaces. Although not common, we have recorded
specimens from Darwin, Genovesa, Floreana (Devil's Crown), Santa Fe (Roca
Pequino), and San Cristobal (Point Pitt).

Orange chromotype of *Muricea* sp. 2.

Multicolored colonies of *Muricea* Sp. 2 are relatively common at Darwin Island. At right (inset) are branches of a white color morph with polyps withdrawn.

Multiple colonies beneath an overhang at Genovesa.

108

Heterogorgia hickmani Breedy & Guzman, 2004. Colonies consist of numerous mostly single branches arising from a continuous encrusting holdfast. The bright yellow or lemon colored polyps extend vividly from the branches and usually are extended during both day and night hours. More robust branches and more crowded spacing of the polyps in *H. hickmani* distinguish it from *H. verrucosa*. **Habitat & range:** On rock substrates and observed from 6 to 11 m depth. Confirmed sightings at Isabela, Floreana, Isla Tortuga, Santa Fe, Rabida, Gordon Rocks, Española, San Cristobal (Breedy, et al. 2008). Probably present throughout the central archipelago. Present known distribution Galápagos Islands.

Left, Branch end of *Heterogorgia hickmani* with fully-extended polyps.

Heterogorgia verrucosa Verrill, 1868. Colonies consist of numerous branches that divide dichotomously, with polyps that extend prominently from the branches. The coenenchyme is yellowish brown, the polyps yellow. This species differs from the more common *H. hickmani* in having more subdivided and less robust branches, and greater spacing between polyps (20 polyps/cm as compared to 40-50 polyps/cm in *H. hickmani*). **Habitat & range:** The single specimen of this species from Galápagos was collected at Gordon Rocks at 23 m. The species is reported from Costa Rica, Panama, and Pacific Colombia.

Closeup of the polyps of
Heterogorgia verrucosa.

Class Anthozoa, Subclass Octocorallia, Order Gorgonacea,
Family Gorgoniidae

Angel Chiriboga

Darwin Bay, Genovesa

Eugorgia daniana Verrill, 1868. Branching spread in one plane, each axis giving off lateral branches with free ends. The branches do not anastomose with neighboring branches into a network, a characteristic that distinguishes this genus from the genus *Pacifigorgia.* Color brown, tan, yellow, or red. The spicules are mostly disc spindles. **Habitat & range:** From Darwin Bay, Genovesa, 30 m on edge of wall; also Pinzon and Sin Nombre Island. This species is reported from Costa Rica and Panama.

Detail of web structure
of *Eugorgia daniana.*
The calices are yellow or
red on yellow to orange
coenenchyme.

Leptogorgia alba (Duchassaing & Michelotti, 1864). This handsome gorgonian is almost pure white in color, although bearing a yellowish or reddish cast when the red polyps are extended. The trunk gives rise to several main branches, subdividing into branchlets that may subdivide again. The branches do not coalesce but are free, slender, and whiplike. Each major branch of the colony grows mostly in one plane but the colony as a whole consists of branches oriented in different directions. Specimens in Galápagos are usually small, 10 to 20 cm in height. Spicules are spindles, the shorter ones with warts on one side fused like those of disc spindles; long spindles are symmetrical or with simple and conical warts on one side, elsewhere compound. **Habitat & range:** Darwin and Wolf Islands, especially common at Darwin's Arch, Darwin Island, not reported elsewhere in Galápagos. The species is recorded from El Salvador south to Panama and the Galápagos Islands.

Left, Detail of *Leptogorgia alba* colony showing slightly prominent red polyp mounds. *Above,* With red polyps extended.

Angel Chiriboga

Wolf, with blacktip cardinalfish

Pacifigorgia dampieri Williams & Breedy, 2004. Fan single in one plane or with several secondary fans in different planes, these sometimes lying perpendicular to the main fan. Intricate network of open meshes averaging about 6 meshes/cm^2 and supported by prominent yellowish orange-colored midribs, which extend from a strong, orange-colored encrusting holdfast. The milky white polyps extend from small, oval, slightly raised mounds, brick red to dull orange in color. The coenenchyme is brick red to rust orange in color. **Habitat & range:** On rocky substrates from 9 to 21 m depth, apparently restricted to the islands of Darwin, Wolf, and Roca Redonda. Endemic to the Galápagos Islands.

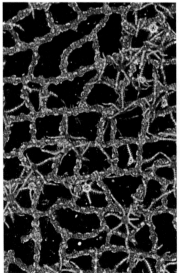

Left, Detail of network with polyps retracted, showing the brick red coenenchyme with dull orange, oval polyp mounds. These colors are diagnostic for the species. *Above,* Network with white polyps extended. The branches bear numerous epizoic brittle stars (*Ophiothela mirabilis*) that compete with the gorgonian polyps for planktonic food carried through by the current.

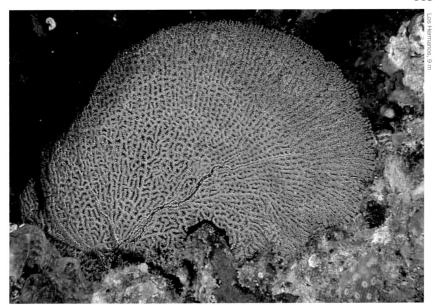

Los Hermanos, 9 m

Pacifigorgia darwinii (Hickson, 1928). Colonies with fans branching in one plane, usually two or three more or less kidney-shaped fans from a single holdfast. Meshes form a finely reticulated network, approximately 7 meshes/cm². Supporting midribs are not prominent. Polyps are white to yellow, arising from oval, mound-like protruberances. The retracted polyp mounds are yellow to cream to white and the coenenchyme is dark reddish purple. This is the most common of *Pacifigorgia* species in the central archipelago. **Habitat & range:** Firmly attached to solid substrates, 3 to 26 m depth. Western and central archipelago from Fernandina to San Cristobal. This species does not occur in the northwest islands of Darwin and Wolf. Endemic to the Galápagos Islands.

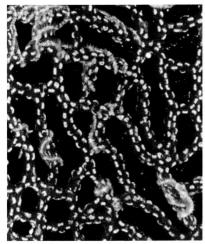

Left, Detail of branching pattern in living P. darwinii with retracted polyps. Polyp mounds are white or yellow on purple to reddish purple coenenchyme. *Above,* Colony with polyps extended.

114

Angel Chiriboga

Genovesa, 10.5 m

Pacifigorgia rubripunctata Williams & Breedy, 2004. Single primary fan, often with 2-3 secondary fans that protrude at right angles to main fan. Network density averages about 6 meshes/cm². No distinct midribs cross the fan. The polyps are white. Four rows of polyp mounds, orange in color, are distinctly raised and arranged in two longitudinal rows on each face of the fan. The orange polyp mounds contrast sharply with the darker coenenchyme, which varies in color from dark reddish-purple to reddish-yellow. **Habitat & range:** Attached firmly to rock, 3-20 m depth. This species occurs in the central archipelago, including Santiago, Rabida, Santa Cruz, Española, Floreana, Los Hermanos, and Roca Redonda. Endemic to the Galápagos Islands.

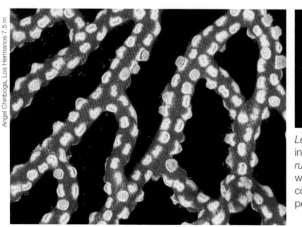

Angel Chiriboga, Los Hermanos 7.5 m

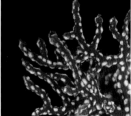

Left, Detail of branching pattern of living *P. rubripunctata* with polyps withdrawn. *Above*, Young colony with withdrawn polyps.

Pacifigorgia symbiotica Williams & Breedy, 2004. Colony of a single wide, stiff fan, or with secondary fans. The network is formed of relatively thick mesh branches (1.5 to 2.5 mm in width) with about 6-8 meshes/cm². The bright yellow polyp mounds cover most of the surface of the mesh branches and are responsible for the overall yellow color of living colonies. The polyps are white but withdrawn during daylight hours. The coenenchyme, mostly hidden by the polyp mounds, is deep reddish purple. **Habitat & range:** Firmly attached to rock substrates. Range, northern islands of Darwin and Wolf, and Isabela (Caleta Iguana and Point Rosa). Endemic to the Galápagos Islands.

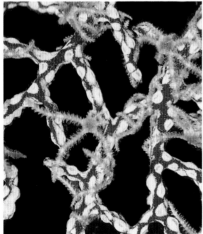

Left, Detail of the *P. symbiotica* with epizoic brittle stars, *Ophiothela mirabilis.* The gorgonian also hosts a caridean shrimp and ovulid snail that mimic the gorgonian's color patterns (Williams & Breedy, 2004). *Above,* Colony with polyps extended.

Angel Chiriboga

Sin Nombre, 21 m

Pacifigorgia sp. 1 The colony is distinguished by a widely-branching fan with bright red coenenchyme. Prominent white midribs extend through the fan from the holdfast. The polyps are white, extending from mounds, which appear to be the red color of the coenenchyme. Of the four described *Pacifigorgia* species, *P. rubripunctata* most closely resembles this new species, but the coenenchyme is much brighter red than that of *P. rubripunctata*. **Habitat & range:** First recorded by Angel Chiriboga at Sin Nombre in 2004 and again at Sin Nombre in 2006 by the author. Range unknown.

Class Anthozoa, Subclass Octocorallia, Order Pennatulacea,
Family Virgulariidae

Joshua Feingold

Devil's Crown, Floreana, 9 m

Paul Humann

***Scytalium* sp.** The colony is bilaterally symmetrical throughout the length of the rachis, which bears thin, fleshy polyp leaves with broadest leaves at the base of the rachis. The polyps are tubular with the distal portions (anthocodiae) retractile. **Habitat & range:** Sandy substrates. In Galápagos the genus is known from a single photograph (above) taken by Joshua Feingold at Devil's Crown in 2003. Described species of the genus are distributed in the Indo-Pacific.

Virgularia galapagensis Hickson, 1930. Long, slender, unbranched, bilateral colony. From the central stalk, or rachis, extend lateral rows of "leaves," each of which bears numerous white, retractile, polyps. Spicules absent except for small platelets in stalk interior. **Habitat & range:** Sandy substrates, such as the sand bottom at Tagus Cove, where this species was once abundant. It was recorded by the CDRS ROV in 2005 at 48 m offshore from Bartolomé on sandy bottom, where it may be grazed by sea turtles. The author collected a single specimen at James Bay in 2004. Probably endemic to Galápagos.

118

Graham Edgar

Wolf Island, 38 m, May 2006

Ptilosarcus undulatus (Verrill, 1865). Bilateral colony with well-developed polyp leaves
bearing one or more marginal rows of polyps.
Spicules are minute oval bodies, plates, rods,
and prismatic needles. **Habitat & range:** Sand
and mud substrates. This species is considered
vulnerable to extinction. The discovery in 2006
of specimens at 38 m at Wolf Island were the first
observed in the archipelago since the 1982-83 El
Niño event. The species occurs from the Gulf of
California to Peru and the Galápagos Islands. It is
very close in form to *Ptilosarcus gurneyi,* which
ranges from Alaska to southern California.

Class Anthozoa, Subclass Octocorallia, Order Pennatulacea,
Family Veretillidae

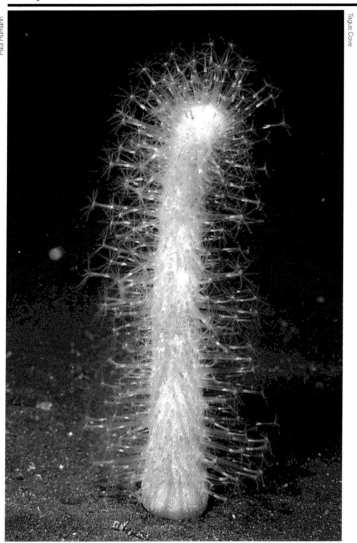

Cavernulina cf. *darwini* Hickson, 1921. Colony of polyps arising separately from club-shaped rachis. No trace of bilaterality. Spicules of oval or elongated rods. Species identification is tentative because the photographed specimen was not collected for examination. **Habitat & range:** Sand and mud-sand substrates. Presumably endemic to Galápagos. The single specimen on which Hickson derived the species' description was part of Charles Darwin's Beagle collection, collected from San Cristobal in 1835. It had not been seen since the 1982-83 El Niño event until recorded in April of 2005 by ROV at 48 m offshore from Bartolomé (S. Banks, pers. comm.)

The delicate, plume-like branching of a colonial hydroid, suggesting a harmless plant, belies its predaceous nature: feathery plumes bear thousands of stinging tentacles lying in wait for passing prey. Most hydroids are harmless to humans, although one common Galápagos species, *Macrorhynchia philippina*, is appropriately known as the "stinging hydroid." Another mildly venomous species, *Pennaria disticha,* is abundant and widespread in shallow waters throughout the region. Of the 96 named species of hydroids from the Galápagos Islands (Calder, et al., 2003) most are too small and inconspicuous to be recognized as among the most abundant of seashore animal life. Although hydroids occur almost everywhere, only a few are more than a few millimeters in height and of interest to divers and snorkelers. Even these larger forms are often mistaken for seaweeds. Unfortunately for casual divers, the full beauty of hydroids cannot be appreciated until viewed through a microscope. Then is revealed the elegance of row upon row of glassy, chalice-like cups, each with an emerging graceful circle of transparent tentacles.

Hydroids are divided into two main groups according to the extent that polyps are protected by a horny covering. Most familiar are the **thecate** hydroids (from the L. *theca*, box), fully sheathed hydroids, such as species of the genus *Obelia*. Thecate hydroids usually have branching stems with tentacled polyps housed in protective, goblet-shaped transparent cups. In **athecate** hydroids, the horny cup is absent (or scanty) and the polyps are naked. One of the most common of Galápagos naked hydroids is *Pennaria disticha*, found throughout the archipelago.

The author expresses his appreciation to Dale R. Calder, who accompanied the author on one research trip to Galápagos, and who made the taxonomic determinations in this section.

122

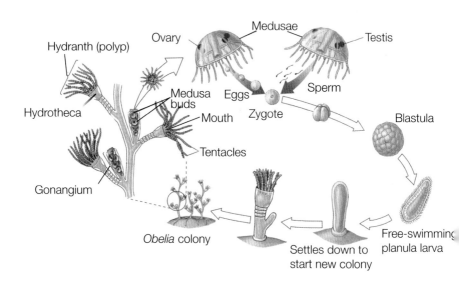

Life cycle of a hydroid, *Obelia,* showing alternation of polyp (asexual) and medusa (sexual) stages. *Obelia* is an example of a thecate hydroid; in these the polyp is housed in a protective cup, the hydrotheca, into which the polyp can withdraw for protection.

Phylum Cnidaria, Class Hydrozoa, Order Anthoathecatae,
Family Tubulariidae

Ectopleura integra (Fraser, 1938) (Syn.: *Tubularia integra*). Colonies forming clumps of flower-like, athecate hydranths on long tube-like stems; each hydranth with an aboral whorl of large tentacles and an oral whorl of smaller tentacles; gonophores with medusa buds with four small tentacles developed before release. **Habitat & range:** Subtidal on firm substrates, throughout the archipelago. Tropical eastern Pacific.

View of complete colony of *Ectopleura integra*.

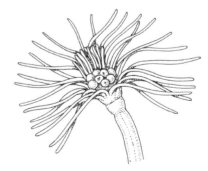

Hydranth of *Ectopleura integra*, showing medusa buds at base of the manubrium.

124

Ectopleura media Fraser, 1948. Colony similar to *Ectopleura integra* with long tube-like stems; each hydranth with an aboral whorl of large tentacles and an oral whorl of smaller tentacles. However, the gonophores of *E. media* have no tentacles and are never liberated (rather than gonophores that have four tentacles that are liberated as free medusae, as in *E. integra*). The manubrium is longer in *E. media* than in *E. integra*. **Habitat & range:** Subtidal on firm substrates, throughout the archipelago. Believed endemic to Galápagos.

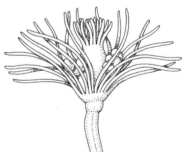

Ectopleura media, showing elongate manubrium (not an invariable feature) and gonophores that remain attached.

Closeup view of hydranths of *Ectopleura media*.

Phylum Cnidaria, Class Hydrozoa, Order Anthoathecatae,
Family Pennariidae

Pennaria disticha Goldfuss, 1820 (Syn.: *Halocordyle disticha*). Colonies large, conspicuous, often four or more cm high, with alternate branches each with a terminal hydranth. Hydranths athecate, with an aboral whorl of large thread-like tentacles and several oral whorls of smaller knobbed tentacles. A venomous species. **Habitat & range:** Common throughout the archipelago. Specimens from sheltered waters are larger and more gracile than wave-swept specimens, which tend to be small and compact. Circumglobal in tropical, subtropical, and warm-temperate waters.

Closeup view of the hydranths of *Pennaria disticha*.

Pennaria disticha, showing athecate hydranths on alternate branches.

Phylum Cnidaria, Class Hydrozoa, Order Anthoathecatae,
Family Eudendriidae

Eudendrium carneum Clarke, 1882. Colonies bushy, extensively branched, to 4 cm or more high; branches with terminal hydranths. Hydranths athecate, hypostome flared and often trumpet-shaped; tentacles thread-like, in one whorl, fewer than 35. **Habitat & range:** Subtidal on rocks. The genus is cosmopolitan.

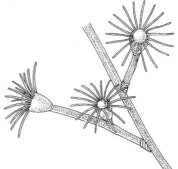

Branching stem of *Eudendrium carneum* with three athecate and urn-shaped hydranths.

Colony of *Eudendrium carneum* showing typical bushy growth form.

Phylum Cnidaria, Class Hydrozoa, Order Leptothecatae,
Family Aglaopheniidae

Aglaophenia diegensis Torrey, 1902. Colonies are fern-like, with mono-podal feather-like plumes 2-4 cm or more high, with alternating branchlets. The hydrothecae are sessile, relatively shallow (less than twice margin diameter), margin with nine distinct cusps. Other species of the genus *Aglaophenia* have been reported from Galápagos, but are considered uncommon or of doubtful identity. **Habitat & range:** Common throughout the archipelago in protected, often hidden, subtidal substrates. Tropical and temperate eastern Pacific.

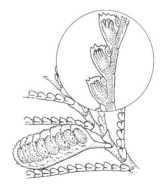

Left, The grain-like structures on the branches of this colony of *A. diegensis* are corbulae, protective pods covering the reproductive gonangia. *Above,* Branch showing thecate hydrothecae and one corbula at left.

Macrorhynchia philippina Kirchenpauer, 1872. Colonies of this stinging hydroid are large, plumose, extensively branched, 5-10 cm high; ultimate
branches alternate, appearing white underwater; hydrothecae sessile, slip-
per-shaped, with prominent internal septum. A venomous species. **Habitat
& range:** Common throughout the archipelago. Tolerant of strong current.
Circumglobal in tropical, subtropical, and warm-temperate waters.

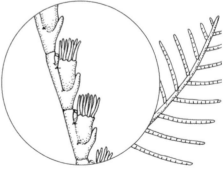

Left, Closeup of *M. philippina*.
"Ostrich-plume" hydroid with featherlike growth
form. *Above,* Branch with inset showing
hydrothecae.

Phylum Cnidaria, Class Hydrozoa, Order Leptothecatae,
Family Haleciidae

Nemalecium lighti (Hargitt, 1924). Hydroids up to 2 cm high. Hydrothecae very shallow, saucer-shaped, not enclosing hydranths. Hydranths elongate,
constricted just below whorl of tentacles. Tentacles of two types: most long
and thread-like but usually with two thickened and finger-shaped "nematodac-
tyls" that curve over distal end of hydranth. **Habitat & range:** Subtidal in areas
of strong current. Circumglobal distribution. This hydroid, a new record for
Galápagos Islands, was collected from Wolf Island, where it is common.

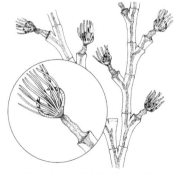

Left, Nemalecium lighti. Above, Showing
elongate hydranths in shallow hydrothecae.

Phylum Cnidaria, Class Hydrozoa, Order Leptothecatae,
Family Campanulariidae

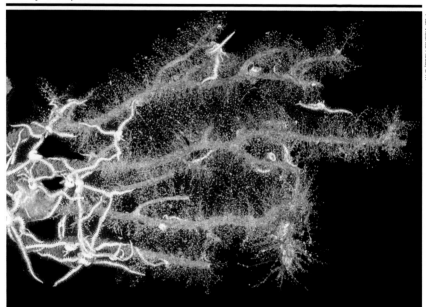

Obelia dichotoma (Linnaeus, 1758). Colonies small, 1-2 cm high, with zig-zag stems; hydrothecae pedicellate, tumbler-shaped, with smooth margin; hydranths with knob-shaped manubrium; gonophores liberated as free medusae. **Habitat & range:** This species is widely distributed in Galápagos. It is a morphologically variable species, whose growth form is affected by current, more compact in strong current. Reported to be essentially cosmopolitan but the taxonomy of the genus is in need of reexamination (D. Calder, pers. comm.).

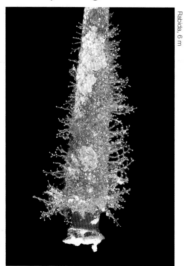

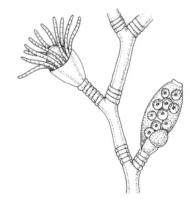

Left, Obelia dichotoma on a pencil urchin spine. *Above, Obelia dichotoma,* showing a hydranth and a gonotheca containing gonophores on zig-zag stem.

Phylum Cnidaria, Class Hydrozoa, Order Leptothecatae,
Family Sertulariidae

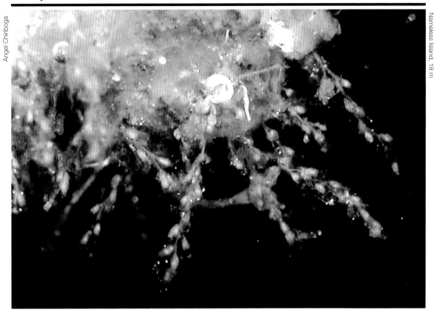

Angel Chiriboga

Nameless Island, 18 m

Sertularella ampullacea Fraser, 1938. Hydroids tiny, less than 1 cm high.
Hydrothecae appearing to have a pedicel but actually sessile on stem, elongate-
oval, with wavy walls, margin with four cusps; operculum of four valves. **Habitat & range:** Subtidal on semicryptic habitats in the central archipelago. Eastern
Pacific from Gulf of California to Colombia and Galápagos Islands.

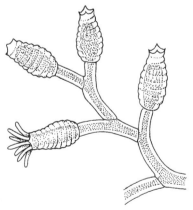

Left, Sertularella ampullacea.
Above, S. ampullacea with elongate
hydrothecae with wavy walls.

Los Hermanos, 11.5 m

Tridentata turbinata (Lamouroux, 1816). Colonies about 1-2 cm high, with erect, unbranched stems bearing two longitudinal rows of hydrothecae in opposite pairs. Hydrothecae sac-shaped, borne directly on stem, each bent abruptly outwards, with a horseshoe-shaped ridge of perisarc marking the bend internally. Hydrothecal orifice with three triangular cusps, two prominent and one inconspicuous; no cusps borne within the hydrothecal cavity. **Habitat & range:** Subtidal on rock. Circumglobal in tropical and subtropical waters.

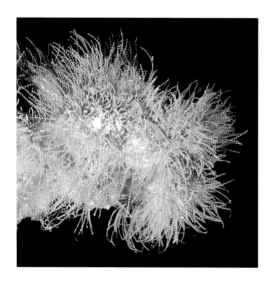

Left, *Tridentata turbinata,* colony of unbranched stems.
Above, Showing opposite pairs of hydrothecae on unbranched stem.

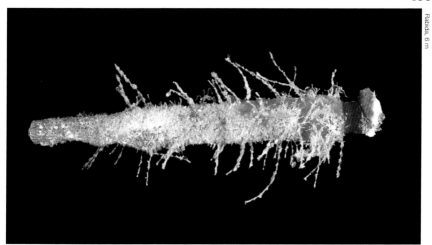

Rabida, 6 m

Dynamena quadridentata (Ellis & Solander, 1786). Colonies small; 1-2 cm high, unbranched; hydrothecae sessile, horn-shaped, occurring in close groups of 2-7 or more closely appressed pairs, margin with two prominent cusps. Colonies in photograph are growing on a slate pencil urchin spine. **Habitat & range:** Central archipelago, probably also in western archipelago. Circumglobal in tropical and subtropical waters.

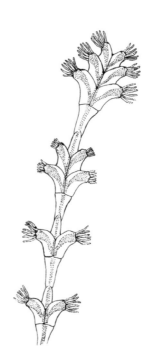

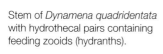

Stem of *Dynamena quadridentata* with hydrothecal pairs containing feeding zooids (hydranths).

Phylum Cnidaria, Class Hydrozoa, Order Leptothecatae,
Family Plumulariidae

Tagus Cove, 12 m

Plumularia floridana Nutting, 1900. Colonies small, fern-like, to 1.5 cm
high; branches alternate, unbranched, to 3.2 mm long; hydrothecae sessile, cup-
shaped, on branches only and not on stem, margin smooth; gonophores small,
globular, in axis between stem and branches. **Habitat & range:** Subtidal on firm
substrates. Form variable, being larger in gentle current, more compact in strong
current. This is a eurytopic species that is tolerant of reduced salinities. Presumed
to be circumglobal in tropical waters.

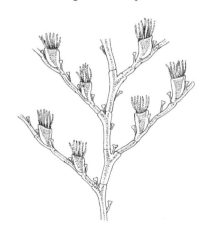

Plumularia floridana, showing alternate branches with hydrothecae.

Addendum to Published Field Guides in the Galápagos Marine Life Series.

The species described in this section are echinoderms and opisthobranch molluscs collected and identified after publication of *A Field Guide to Sea Stars and other Echinoderms of Galápagos* and *A Field Guide to Marine Molluscs of Galápagos.*

Phylum Echinodermata, Class Asteroidea, Order Paxillosida, Family Luidiidae

Luidia superba A.H. Clark, 1917. Dramatically large sea star, reaching 1 m in diameter, probably the largest five-armed asteroid in the world. Stout arms that taper to a blunt end are covered with large paxillae, each paxilla a skeletal plate that supports pillars bearing a crown of small spinelets. Color grey to brown. The holotype was collected in 1888 by dredging off the coast of Colombia. In Galápagos, specimens have been collected from Tagus Cove and additional sightings made by the DeRoy family at Stephens Bay, San Cristobal, Conway Bay, Isabela, and Academy Bay, Santa Cruz (Downey & Wellington, 1978). **Habitat & range:** Sand, occasionally rocky substrates. Wellington's observations suggest the star is active at night but lies buried under the sand at 10-12 cm depth during the day. Colombia, Ecuador, and Galápagos Islands.

136

Luidia bellonae Lütken, 1864. This photograph of a living specimen supplements the preserved specimen pictured in *A Field Guide to Sea Stars and other Echinoderms of Galápagos,* p. 7.

Luidia columbia (Gray, 1840). This sea star, mistakenly identified as *Luidia foliolata* Grube, 1866, on page 6, *A Field Guide to Sea Stars and other Echinoderms of Galápagos*, is correctly identified as *Luidia columbia* (Gray, 1840). Chris Mah, who brought the error to my attention, writes "The two are almost exactly alike... the only difference is that the spines in *L. columbi*a are very slender and more needle-like compared to those in *L. foliolata*, which are thicker and more tooth-shaped."

Phylum Echinodermata, Class Asteroidea, Order Paxillosida,
Family Astropectinidae

Tethyaster canaliculatus (A.H. Clark, 1916). Large sea star with large
disc and five broad, flattened arms, often unequal in length. The tube feet lack
suckers. Marginal arm plates large and conspicuous. Paxilla closely spaced,
uniform in size. Diameter to 50 cm. Seldom seen and considered rare. Like
Luidea superba, it forages at night and burrows beneath the surface of sand
during the day. **Habitat & range:** On subtidal sand, often burrowing beneath
the surface. Baja California to Panama and Galápagos Islands (new record).

*Tethyaster
canaliculatus,*
photographed at
Punta Espinosa,
Fernandina during
a night dive.

Phylum Echinodermata, Class Asteroidea, Order Forcipulatida,
Family Asteriidae

Rabida

Coronaster marchenus Ziesenhenne, 1942. An unusual sea star, first
collected at Marchenus Island, Galápagos. With 9 to 17 weak arms, each bearing
four rows of large, tapering blunt-tipped spines. In life, the ends of lateral spines
are enclosed by a conspicuous fleshy bundle of minute crossed pedicellaria; the
bulbs tend to collapse and mostly disappear in preserved specimens. There are
two rows of large tube feet. Arms up to 8 cm long. The sea star may reproduce
by fission by casting off a portion of the disk with one or more arms, additional
arms then regenerating on the new individual. Individuals consequently bear
arms of varying length. **Habitat & range:** Rocky substrates, 6 to 91 m depth.
We have collected specimens from Pinzon, Rabida and Floreana and it probably
exists throughout the archipelago. Cocos and Galápagos Islands.

Phylum Echinodermata, Class Asteroidea, Order Valvatida,
Family Asterodiscididae

Paulia horrida Gray, 1840. Five-armed sea star with numerous conical, sharp-pointed spines on the dorsal and lateral surfaces of the arms and large central disc. Dorsal surface flat; marginal plates largely concealed. Diameter to 15 cm. **Habitat & range:** On rocky substrates. Ecuador to Peru and Clarion, Cocos and Galápagos Islands.

Paulia horrida,
color variant.

Phylum Echinodermata, Class Asteroidea, Order Valvatida,
Family Ophidiasteridae

Leiaster teres (Verrill, 1871). This color variant of the species described
on p. 10 of *A Field Guide to Sea Stars and other Echinoderms of Galápagos* is
occasionally seen at the northern islands of Darwin and Wolf.

Phylum Echinodermata, Class Holothuroidea, Order Apodida, Family Synaptidae

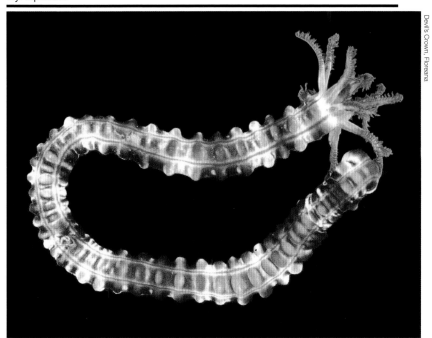

Euapta godeffroyi (Semper, 1868). This remarkable sea cucumber has a soft, highly extensible translucent body with brown bands and brown or white lateral tubercles. Featherlike tentacles surround the mouth. It lacks tube feet and is sticky to the touch. Active mainly at night. **Habitat & range:** Found on sand bottom in shallow water. Range Indian Ocean to the Eastern Pacific from Mexico to Panama and Galápagos Islands (new record).

142

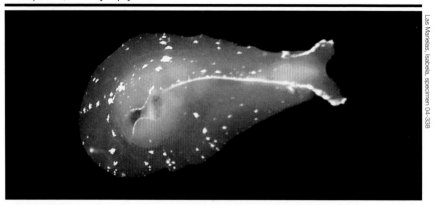

Las Marielas, Isabela, specimen 04-338

Phyllaplysia padinae Williams & Gosliner, 1973. An olive-green to
brownish green sea slug with random white spotting on the dorsal surface. The
brownish-green ground color is similar to *Padina* sp, the brown alga host.
The rhinophores are short with noticeably enlarged tips. Length to 4.5 cm. In
the original description (Williams & Gosliner, 1973) the species is described as
moving by inch-worm-like locomotion. It feeds by scraping epiphytic diatoms
and epidermal cells from the host algae. **Habitat & range:** Inhabits *Padinae*
algae, apparently exclusively. Present known range is Gulf of California and
Galápagos Islands (new record).

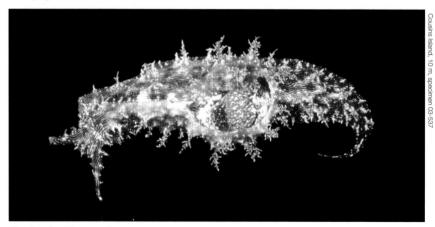

Cousins Island, 10 m, specimen 03-537

Stylocheilus striatus (Quoy & Gaimand, 1824) (formerly *Stylocheilus
longicauda*). This common sea hare, widely distributed in both the east and
west Pacific as well as the Caribbean, appeared on p. 117 of *A Field Guide to
the Marine Molluscs of Galápagos*. The name was changed from *S. longicauda*
to *S. stiratus* in 1999 by Bill Rudman (see http://www.seaslugforum.net/
factsheet.cfm?base=stylnome).

The author is indebted to Terry Gosliner for identification of opisthobranch molluscs
described in this section.

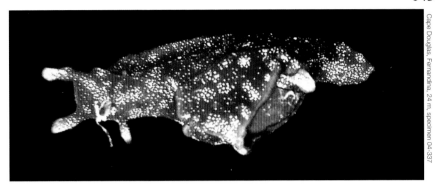

Aplysia parvula Mörch, 1863. Small sea hare, less than 6 cm in length, with large oral tentacles anteriorly and two rhinophores just above the eyes. The black-edged parapodia flaps surround an opening to a cavity enclosing a small, plate-like shell. Color variable, depending upon the seaweed on which it is feeding. Often dark maroon body mottled with white spots, black trim on parapodia. **Habitat & range:** Feeds on folios red algae. The species is widespread circumtropical.

Phylum Mollusca, Class Gastropoda, Subclass Opisthobranchia, Order Notaspidea, Family Pleurobranchidae

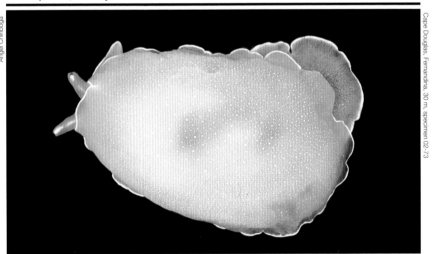

Berthella californica (Dall, 1900). Translucent white side-gilled opistho-branch with white spots on the mantle, oral veil and foot. The species is distinguished by an opaque white line on the sheetlike extension of the mantle that overhangs the foot, on the foot itself and on the rhinophores. Size reported to 8 cm in length, but Galápagos specimens are less than half this length. **Habitat & range:** On rocky substrates. In California the species is reported to feed on ascidians. Alaska to Panama and the Galápagos Islands.

144

Phylum Mollusca, Class Gastropoda, Subclass Opisthobranchia, Order
Nudibranchia, Family Chromodorididae

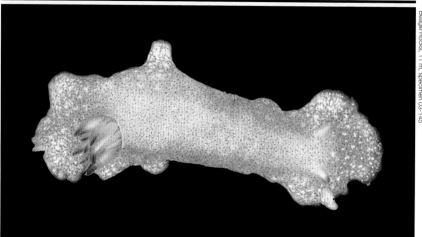

Beagle Rocks, 11 m, specimen 03-145

Chromodoris baumanni Bertsch, 1970. Color morph of the species
described on p. 123 of *A Field Guide to Marine Molluscs of Galápago*s, seen
throughout the archipelago.

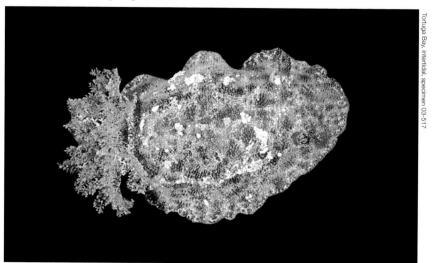

Tortuga Bay, intertidal, specimen 03-517

***Otinodoris* sp**. A large, uncommon dorid that may reach 15 cm in length,
with a furry appearance caused by papillae that completely cover the mantle.
The largest papillae are elevated into small mounds. The mantle margin has
regular undulations. Body color fawn brown with mounds and papillae varying
in color from near-white to dark brown. **Habitat & range:** Low intertidal. The
genus is known from Australia, but full geographic distribution is unknown.
New record for Galápagos.

Phylum Mollusca, Class Gastropoda, Subclass Opisthobranchia, Order
Nudibranchia, Family Platydorididae

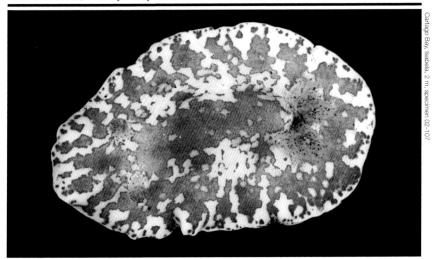

Cartago Bay, Isabela, 2 m, specimen 02-107

Platydoris carolynae Mulliner & Sphon, 1974. A leathery, flattened dorid.
We have found this color variant to be more commonly encountered than the
black and white specimen pictured on p. 124 of *A Field Guide to the Marine
Molluscs of Galápagos.*

Phylum Mollusca, Class Gastropoda, Subclass Opisthobranchia, Order
Nudibranchia, Suborder Doridacea, Family Dendrodorididae

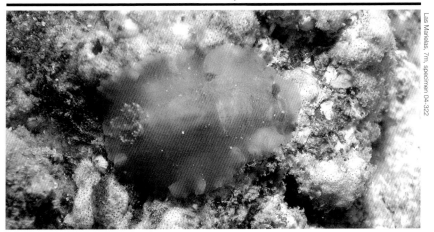

Las Marielas, 7m, specimen 04-322

Dendrodoris fumata (Rüppell & Leuckart, 1831). Body flattened, soft with
broad mantle highly ruffled on the edges. The small head bears two bulbous
rhinophores. Color variable but typically reddish-orange, becoming darker with
age. Surface covered with low papillae. Feeds principally on sponges. Length
to about 4 cm. **Habitat & range:** On rocky substrates with sponges on which it
feeds. Widespread throughout the tropical Indo-Pacific and on the Pacific coast
of Mexico and Costa Rica, and the Galápagos Islands.

Phylum Mollusca, Class Gastropoda, Subclass Opisthobranchia, Order
Nudibranchia, Suborder Doridacea, Family Discodorididae

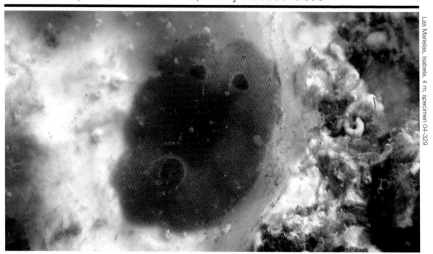

Thordisa cf. *rubescens* Behrens & Henderson, 1981. A cryptobranch dorid,
red-orange in color with gold encrustations across the mantle. Inflated papillae
give the dorsal surface a velvety texture. Rhinophores and gills are tipped with
brown. **Habitat & range:** The single Galápagos specimen found under a rock
at Las Marielas is identical or very close to *Thordisa rubescens,* known from
California.

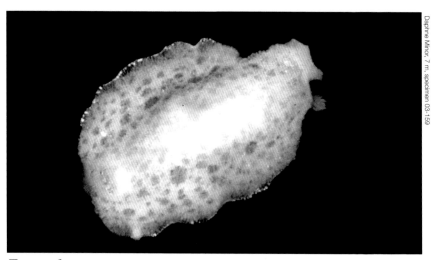

Tayuva ketos Marcus & Marcus, 1967. A white to tan dorid, mottled with
brown spots and light brown rhinophores. The visceral hump is mottled with
very small tubercles. When secured to a rock, the animal resembled an ascidian.
Habitat & range: Intertidal and shallow subtidal, usually under rocks. Northern
Mexico (Gulf of California) and Galápagos Islands (new record).

Unknown dorid. Light red dorid with crimson red gills, tubercles and rhinophores. **Habitat & range**: Sublittoral. Geographic distribution unknown. All specimens we have observed have been on a white sponge, *Chalinula* sp. We have observed this dorid at Pinzon, Devil's Crown, and Guy Fawkes.

148

Phylum Mollusca, Class Gastropoda, Subclass Opisthobranchia, Order
Nudibranchia, Suborder Aeolidina, Family Fionidae

Fiona pinnata (Eschscholtz, 1831). Color cream to brown and variable,
dependent upon diet which is mainly goose barnacles, especially of the genus
Lepas. The cerata on the lateral surfaces are uniquely sail-shaped; the dorsal
midsection is clear of cerata. Length to 20 mm. **Habitat & range**: This species
is almost always found on floating objects such as driftwood, kelp rafts, fishing boats and buoys to which are attached pelagic barnacles (*Lepas* sp.). Range
cosmopolitan.

Phylum Mollusca, Class Gastropoda, Subclass Opisthobranchia, Order
Nudibranchia, Suborder Aeolidina, Family Aeolidiidae

Favorinus elenalexiae Garcia & Troncoso, 2001. This aeolid, pictured and
described as *Favorinus* sp. on p. 128 of *A Field Guide to Marine Molluscs of
Galápagos* has now been described to species.

Phylum Mollusca, Class Gastropoda, Subclass Opisthobranchia, Order
Nudibranchia, Family Scyllaeidae

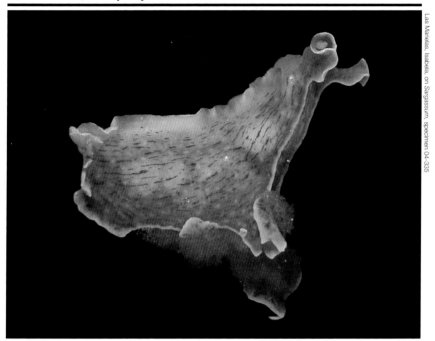

Crosslandia daedali Poorman & Mulliner, 1981. This species has two
large, wing-like lobes on the parapodia, with the anterior lobe larger than the
posterior. Along the sides and beneath are fine chestnut-brown longitudinal
lines. On some specimens are small blue spots between the parapodia. Along
the side of the body are 4 to 6 conical tubercles. Length to 3 cm. **Habitat &
range:** This species is always found with brown algae which it mimics in color.
It is common on *Sargassum*. It feeds on hydroids growing on the algae. Range
previously reported as Gulf of California to Costa Rica. This is the first record of
the species south of Costa Rica.

Glossary

acontia (sing. **acontium**). Threadlike structures bearing
nematocysts located on the mesenteries of sea anemones.

acrorhagi. Marginal tubercles on some sea anemones that contain
nematocysts.

ahermatypic. Literally "without reef-building," (see hermatypic),
and applied to corals that lack zooxanthellae and do not
contribute to reef-building.

athecate. Hydroid with naked polyps, not protected by a
hydrotheca.

autotroph. An organism that builds its organic nutrients from
inorganic raw materials. Self-nourishing.

azooxanthellate. Lacking zooxanthellae.

basaltic. Of basalt, an igneous rock of volcanic origin.

brachycnemic. A feature used in the classification of zoanthids
in which the fifth pair of mesenteries counting from the dorsal
directives are incomplete. The contrasting condition—the fifth
pair of mesenteries are complete—is called macrocnemic.
Zoanthids with incomplete fifth mesenteries belong to the
suborder Brachycnemia; it includes *Zoanthus* and *Palythoa* but
not *Parazoanthus* and *Epizoanthus.*

calices (sing. **calice**). Any of a variety of cup-shaped structures. In
corals, the upper cup-like open end of a corallite. Also spelled
calyx, calyces.

coenosarc (seen'o-sark). In corals, an extension of polyp tissue that
stretches over the surface of the skeleton.

coenenchyme (seen'en-kyme). Mesogleal tissue extending
between the polyps of octocorals, such as gorgonians.

coenosteum (seen-os'te-um). Skeletal material surrounding and
between individual corallites.

collines. Skeletal ridges that separate groups of corallites.

columella. Structure at the central axis of a corallite composed of
the inner ends of septae.

corallite. The skeleton deposited by a single coral polyp.

corallivores. Animals that feed upon corals.

corbulae. Pod-like structure covering and protecting the gonangea of certain hydrozoans.

costae. Skeletal elements that extend outward from the corallite wall.

epizoic. Living on the body of an animal.

fossa. Central depression of a calice, usually containing the columella.

gonagium. Reproductive zooid of a hydroid colony.

gonophore. Sexual reproductive structure developing from reduced medusae in some hydrozoans; it may be retained on the colony or released.

gonotheca. Transparent protective covering of a gonophore.

gorgonin. Scleroprotein in the axial skeleton of gorgonians.

hermatypic. Literally "reef-building," from the Gr, *herma,* reef, and generally applied to coral that contain symbiotic zooxanthellae that may contribute to the building of reefs. However, the term is ambiguous because not all reefs (deepwater reefs, for example) are composed of zooxanthellate corals and not all zooxanthellate corals are found in reefs.

heterotroph. An organism that obtains nourishment from outside sources; includes most animals.

hydranth. Nutritive zooid of a hydrozoan colony.

hydrotheca. Protective cup surrounding the polyp of a hydroid.

hypostome. Oral projection of a hydrozoan.

laminar. Platelike; layered.

manubrium. Conical elevation of hydrozoan polyp or medusa bearing the mouth.

mesentery. Radial partition extending inwards from the wall of the gastrovascular cavity.

nematocyst. Stinging organelle of cnidarians.

perithecal. Surface of the coenosteum between individual corallites.

polyp. Individual of the phylum Cnidaria, usually adapted for attachment to the substratum; often forms colonies.

rachis. Upper polyp-bearing region of a sea pen.

reef. Limestone platform built mainly by corals in tropical seas.

reticulate evolution. Type of evolution in which clades divide to form species and species fuse to form hybrids.

sclerites. Calcareous plates or spicules. In gorgonians, calcareous spicules embedded in the tissue; part of support element.

septa. Radial skeletal elements that extend inward from the corallite wall.

septo-costae. Extensions of the septa that extend outward beyond the corallite. They occur in coral that lack corallite walls and where there is no clear distinction between septa and costae.

siphonoglyph. Ciliated furrow in the gullet of sea anemones.

stolon, stoloniferous. Literally, a shoot, or sucker of a plant. In corals, horizontal outgrowths of polyps from which daughter polyps are budded.

theca. In corals, the wall surrounding an individual corallite.

thecate. Hydroid with tentacled polyps protected by hydrothecae.

verrucae. From the Latin *verruca,* wart. On branches of *Pocillopora* corals, mounds of coenosteum that bear numerous corallites. On sea anemones, wart-like projections of the body wall.

zooxanthellae. Unicellular algae (dinoflagellates) that live symbiotically in the tissues of corals.

Selected References

Corals

Banks, S., M. Vera, and A. Chiriboga. 2008. Characterizing the last remaining reefs: establishing reference points to assess long term change in Galápagos zooxanthellate coral communities. Galápagos Research (in press).

Cairns, S.D. 1991. A revision of the ahermatypic scleractinia of Galápagos and Cocos Islands. Smithsonian Contributions to Zoology, no. 504, 32 pages, 12 plates.

Cairns, S.D. 2001. A generic revision and phylogenetic analysis of the Dendrophylliidae (Cnidaria: Scleractinia). Smithsonian Contributions to Zoology, no. 615, 75 pages, 14 plates.

de Paula, A.F. and J.C. Creed. 2004. Two species of the coral *Tubastraea* (Cnidaria, Scleractinia) in Brazil: a case of accidental introduction. Bulletin of Marine Science, **74**(1):175-183.

Durham, J.W. 1962. Corals from the Galápagos and Cocos Islands. Proceedings of the California Academy of Sciences, 4th series, **32**(2):41-56.

Glynn, P.W. 2003. Coral communities and coral reefs in Ecuador. *In* J. Cortes, ed., Latin American Coral Reefs, pp. 449-472.

Glynn, P.W. and G.M. Wellington. 1983. Corals and coral reefs of the Galápagos Islands. Berkeley, University of California Press, 330 pp.

Glynn, P.W., S.B. Colley, J.H. Ting, J.L. Maté and H.M. Guzmán. 2000. Reef coral reproduction in the eastern Pacific: Costa Rica, Panamá and Galápagos Islands (Ecuador). IV. Agariciidae, recruitment and recovery of *Pavona varians* and *Pavona* sp. Marine Biology, **136**:785-805.

Vera M. and S. Banks. 2008. Health status of the coralline communities of the northern islands; Darwin, Wolf and Marchena of the Galápagos Archipelago. Galápagos Research (in press).

Veron, J.E.N. 1995. Corals in space and time: The biogeography and evolution of the Scleractinia. Comstock/Cornell, Ithaca and London, 321 pp.

Veron, J.E.N. 2000. Corals of the world (3 volumes). Australian Institute of Marine Science, Townsville, Australia.

Wells, J.W. 1983. Annotated list of the scleractinian corals of the Galápagos. *In* P.W. Glynn and G.M. Wellington, Corals and coral reefs of the Galápagos Islands, pp. 212-291, 21 plates.

Sea Anemones and Cerianthids

Carlgren, O. 1949. A survey of the Ptychodactiaria, Corallimorpharia and Actiniaria. Kungliga Svenska Vetenskaps-Akademiens Handlingar, **1**(1):1-121. [*Phellia exlex; Bunodactis* sp*.; Telmatactis panamensis; Antiparactis lineolatus; Calliactis polypus; Aiptasia* sp.]

Carlgren, O. 1950. A revision of some Actiniaria described by A.E. Verrill. Journal of the Washington Academy of Sciences, **40**(1): 22-28. [*Bunodosoma grandis*]

154

Carlgren, O. 1951. The actinian fauna of the Gulf of California. Proceedings of the United States National Museum, 101(3282):415-449. [*Aulactinia* cf. *mexicanum; Telmatactis panamensis*]

Daly, M. and D. G. Fautin. 2004. *Anthopleura mariscali*, a new species of sea anemone (Cnidaria: Anthozoa: Actiniaria) from the Galápagos Islands. Zootaxa, 416:1-8. [*Anthopleura mariscali*]

Dunn, D.F. 1974. Redescription of *Anthopleura nigrescens* (Coelenterata, Actiniaria) from Hawaii. Pacific Science, 28(4):377-382. [*Anthopleura nigrescens*]

Fautin, D.G, C.P. Hickman, Jr., M. Daly and T. Molodtsova. 2007. Shallow-water sea anemones (Cnidaria: Anthozoa: Actiniaria) and tube anemones (Cnidaria: Anthozoa: Ceriantharia) of the Galápagos Islands. Pacific Science, 61(4):549-573. [*Bunodosoma grandis;Phymactis papillosa; Anthopleura nigrescens; Anthopleura mariscali;Telmatactis panamensis; Aiptasia* sp.; *Calliactis polypus; Antiparactis* sp.]

Häussermann, V. 2003. Redescription of *Oulactis concinnata* (Drayton in Dana, 1846) (Cnidaria: Anthozoa: Actiniidae), an actiniid sea anemone from Chile and Perú with special fighting tentacles; with a preliminary revision of the genera with a "frond-like" marginal ruff. Zoologische verhandelingen, Leiden, 345:173-207. [*Actinostella* cf. *bradleyi*]

Häussermann, V. 2004. Re-description of *Phymactis papillosa* (Lesson, 1830) and *Phymanthea pluvia* (Drayton in Dana, 1846) (Cnidaria: Anthozoa), two common actiniid sea anemones from the south east Pacific with a discussion of related genera. Zoologische Mededelingen, Leiden, 78:345-381. [*Phymactis papillosa*]

Mathew, K. 1979. Studies on the biology of a sea anemone, *Anthopleura nigrescens* (Verrill) from the south west coast of India. Bulletin of the Department of Marine Sciences University of Cochin, 10:75-158. [*Anthopleura nigrescens*]

McMurrich, J.P. 1893. Report on the Actiniae collected by the United States Fish Commission Steamer Albatross during the winter 1887-1888. Proceedings of the United States National Museum, 16(930):119-216. [*Actinostella* cf. *bradleyi* (as *Oulactis californica*); *Antiparactis lineolatus* (as *Paractis lineolata*)]

McMurrich, J.P. 1904. The Actiniae of the Plate Collection. Zoologische Jahrbücher, 6 Suppl., no. 2:215-306. [*Antiparactis lineolatus*]

Okey, T.A., S.A. Shepherd, and P.C. Martinez. 2003. A new record of anemone barrens in the Galápagos. Noticias Galápagos, 62:17-20. [*Aiptasia* sp.]

Reimer, A. A. 1973. Feeding behavior of the sea anemone *Calliactis polypus* (Forskål, 1775). Comparative Biochemistry and Physiology, 44A:1289-1301. [*Calliactis polypus*]

Sonnenholzner, J.I., L.B. Ladah, and K.D. Lafferty. 2007. Cascading effects of fishing on Galapagos rocky reef communities. Marine Ecology Progress Series, 343:77-85.

Verrill, A. E. 1869. Review of the corals and polyps of the west coast of America. Transactions of the Connecticut Academy of Arts and Sciences, 1, part 2:377-567. [*Bunodosoma grandis* (as *Cladactis grandis); Phymactis clemattis (papillosa); Antiparactis lineolata (*as *Sagartia lineolata); Telmatactis panamensis (*as *Phellia panamensis); Actinostella bradleyi (*as *Asteractis bradleyi)*]

Cerianthids
den Hartog, J.C. 1977. Description of two new Ceriantharia from the Caribbean region, *Pachycerianthus curacaoensis* n.sp. and *Arachnanthus nocturnus* n.sp., with a discussion of the cnidom and of the classification of the Ceriantharia. Zoologische Mededelingen (Leiden), **51**(14):211-242.

Fautin, G.D., C.P. Hickman, Jr., M. Daly, and T. Molodtsova. 2007. Shallow-water anemones (Cnidaria: Anthozoa: Actiniaria) and tube anemones (Cnidaria: Anthozoa: Ceriantharia) of the Galápagos Islands. Pacific Science, **61**(4):549-573.

Torrey, H.B. and F.L. Kleeberger. 1909. Three species of *Cerianthus* from Southern California. University of California Publications in Zoology, **6**(5):115-125.

Black Coral
Calvopina, M. 2002. Coral negro. *In* Reserva Marina de Galápagos. Linea Base de la Biodiversidad (Danulat, E. & G. Edgar., eds.) p. 389-395. Fundación Charles Darwin/Servicio Parque Nacional Galápagos, Santa Cruz, Galápagos, Ecuador.

Deichmann, E. 1941. Coelenterates collected on the Presidential cruise of 1938. Smithsonian Miscellaneous Collections, **99**(10):14-15.

Martinez, M.H. 1986. Studies on the exploitation of black coral in the Galápagos Islands, Ecuador. *In* G. Davis-Merlen, ed., 1983 Annual Report of the Charles Darwin Research Station, pp., 54-55.

Opresko, D.M. 1976. Redescription of *Antipathes panamensis* Verrill (Coelenterata, Antipatharia). Pacific Science, **30**(23):235-240.

Opresko, D.M. 2001. Revision of the Antipatharia (Cnidaria: Anthozoa), Part I. Establishment of a new family, Myriopathidae. Zoologische Mededelingen, **75**(17):343-370.

Robinson, G. 1982. Investigation of Galápagos Antipatharian corals: Preliminary results. *In* Annual report, pp. 192-208, Charles Darwin Research Station, Galápagos Islands, Ecuador.

Zoanthids
Burnett, W.J., J.A.H. Benzie, J.A. Beardmore and J.S. Ryland. 1997. Zoanthid (Anthozoa, Hexacorallia) from the Great Barrier Reef and Torres Strait, Australia: systematics, evolution and a key to species. Coral Reefs, **16**:55-68.

156

Reimer, J.D. 2006. Latitudinal and intracolony ITS-rDNA sequence variation in the symbiotic dinoflagellate genus *Symbiodinium* (Dinophyceae) in *Zoanthus sansibaricus* (Anthozoa: Hexacorallia). Physiological Research, **54**:122-132.

Reimer, J.D. and C.P. Hickman, Jr. 2008. Preliminary survey of zooxanthellate zoanthids (Cnidaria:Hexacorallia) of the Galápagos and associated symbiotic dinoflagellates (Symbiodinium spp.). Galápagos Research (in press)

Reimer, J.D., S. Ono, A. Iwama, J.Tsukahara, and T. Maruyama. 2006. High levels of morphological variation despite close genetic relatedeness between *Zoanthus* aff. *vietnamensis* and *Zoanthus kuroshio* (Anthozoa: Hexacorallia). Zoological Science (Japan), **23**:755-761.

Reimer, J.D., S. Ono, K. Takishita, J. Tsukahara and T. Maruyama. 2006. Molecular evidence suggesting species in the zoanthid genera *Palythoa* and *Protopalythoa* (Anthozoa: Hexacorallia) are congeneric. Zoological Science (Japan), **23**:87-94.

Reimer, J.D., F. Sinnniger, and C.P. Hickman, Jr. 2008. Zoanthid diversity (Anthozoa:Hexacorallia) in the Galápagos Islands: a molecular examination. Coral Reefs (in press).

Sinniger, F., J.I. Montoya-Burgos, P. Chevaldonné and J. Pawlowski. 2005. Phylogeny of the order Zoantharia (Anthozoa, Hexacorallia) based on the mitochrondrial ribosomal genes. Marine Biology, **147**:1121-1128.

Walsh, G.E. and R.L. Bowers. 1971. A review of Hawaiian zoanthids with descriptions of three new species. Zoological Journal of the Linnean Society, **50**:161-180.

Gorgonians

Bayer, F.M. 1951. A revision of the nomenclature of the Gorgoniidae (Coelenterata: Octocorallia), with an illustrated key to the genera. Washington Academy of Sciences, **41**(3):91-102. [*Pacifigorgia, Eugorgia*]

Breedy, O. and H.M. Guzman. 2005. A new species of alcyonacean octocoral from the Galápagos archipelago. Journal of the Marine Biological Association of the United Kingdom, **85**: 801-807. [*Heterogorgia hickmani; H. verrucosa*]

Breedy, O. and H.M. Guzman. 2007. A revision of the genus *Leptogorgia* Milne Edwards & Haime, 1857 (Coelenterata: Octocorallia: Gorgoniidae) in the eastern Pacific. Zootaxa, **1407**:1-90.

Breedy, O., C.P. Hickman, Jr. and G. Williams. 2008. Octocoral research in the Galápagos Islands. Galápagos Research (in press).

Hickson, M.A. 1928. The Gorgonacea of Panama Bay together with a description of one species from the Galápagos Islands and one from Trinidad. Videnskabelige Meddelelser fra Dansk naturhistorisk Forening i Kobenhavn, **85**:325-422. [*Leptogorgia alba, Pacifigorgia darwinii* (as *Gorgonia Darwinii*)]

Verrill, A.E. 1869. Review of the corals and polyps of the west coast of America. Transactions of the Connecticut Academy of Arts and Sciences **1**(6): 377-558. [*Muricea fruticosa, M. formosa, Heterogorgia verrucosa*]

Williams, G.C. 1995. Living genera of sea pens (Coelenterata: Octocorallia: Pennatulacea): illustrated key and synopses. Zoological Journal of the Linnean Society **113**:93-140.

Williams, G.C. and O. Breedy. 2004. The Panamic gorgonian genus *Pacifigorgia* (Octocorallia: Gorgoniidae) in the Galápagos archipelago, with descriptions of three new species. Proceedings of the California Academy of Sciences **55** (3):55-88. [*Pacifigorgia dampieri; P. darwinii; P. rubripunctata; P. symbiotica*]

Sea Pens

Batie, R.E. 1972. Investigations concerning the taxonomic status of the sea pen *Ptilosarcus gurneyi* (Cnidaria, Pennatulacea). Northwest Science, **46**(4): 290-300. [*Ptilosarcus undulatus*]

Hickson, S.J. 1921. On some Alcyonaria in the Cambridge Museum. Proceedings of the Cambridge Philosophical Society. Mathematical and Physical Sciences **20**(3):366-373. [*Cavernularia darwinii*]

Hickson, S.J. 1930. Some alcyonarians from the Eastern Pacific Ocean. Proceedings of the Zoological Society of London 1930 (14):209-227. [*Virgularia galapagensis*]

Williams, G.C. 1995. Living genera of sea pens (Coelenterata: Octocorallia: Pennatulacea): illustrated key and synopses. Zoological Journal of the Linnean Society **113**:93-140. [*Ptilosarcus, Scytalium, Virgularia, Cavernulina*]

Hydroids

Calder, D.R. 1988. Shallow-water hydroids of Bermuda. The Athecatae. Royal Ontario Museum, Life Sciences Contributions, **148**:1-107. [*Eudendrium carneum*]

Calder, D.R. 1991. Shallow-water hydroids of Bermuda: the Thecatae, exclusive of Plumularioidea. Royal Ontario Museum, Life Sciences Contributions, **154**:1-140. [*Tridentata turbinata, Obelia dichotoma, Nemalecium lighti, Dynamena quadridentata*]

Calder, D.R. 1997. Shallow-water hydroids of Bermuda: superfamily Plumularioidea. Royal Ontario Museum, Life Sciences Contributions, **161**:1-85. [*Plumularia floridana, Macrorhynchia philippina*]

Calder, D.R., J. J.. Mallinson, K. Collins, and C. P. Hickman. 2003. Additions to the hydroids (Cnidaria) of the Galápagos, with a list of species reported from the islands. Journal of Natural History, **37**:1173-1218. [*Tridentata turbinata, Sertularella ampullacea*]

Fraser, C.M. 1938. Hydroids of the 1934 Allan Hancock Pacific Expedition. Allan Hancock Pacific Expeditions, **4**(1):1-74 + 15 plates. [*Sertularella ampullacea, Ectopleura integra (as Tubularia integra)*]

Fraser, C.M. 1948. Hydroids of the Allan Hancock Pacific Expeditions since March, 1938. Allan Hancock Pacific Expeditions, **4**(5):179-335 + 42 plates. [*Aglaophenia diegensis, Ectopleura media*]

Gibbons, M.J. and J.S. Ryland. 1989. Intertidal and shallow water hydroids from Fiji. I, Athecata to Sertulariidae. Memoirs of the Queensland Museum, **27**:377-432. [*Pennaria disticha*]

References for Addendum
(Echinoderms and Opisthobranch Molluscs)

Behrens, D. & R. Henderson. 1981. Two new cryptobranch dorid nudibranchs from California. The Veliger, **24**(2);120-128. [*Thordisa rubescens*]

Clark, A.H. 1916. Six new starfishes from the Gulf of California and adjacent waters. Proceedings of the Biological Society of Washington, **29**:51-62. [*Tethyaster canaliculatus* as *Sideriaster canaliculatus*)

Clark, A.H. 1916. One new starfish and five new brittle stars from the Galapagos Islands. Annals and Magazine of Natural History, (8)17:115-122. [*Tethyaster canaliculatus*)

Clark, A. H. 1917. Two astroradiate echinoderms from the Pacific coast of Colombia and Ecuador. Proceedings of the Biological Society of Washington, **30**:171-174. [*Luidia superba*]

Clark, H. L. 1902. Papers from the Hopkins Stanford Galapagos expedition, 1898-1899. XII. Echinodermata. Proceedings of the Washington Academy of Sciences, **4**:521-531.

Clark, H.L. 1910. The echinoderms of Peru. Bulletin of the Museum of Comparative Zoology at Harvard University, **52**(17)321-358, 14 pls.

Downey, M.E. & G.M. Wellington. 1978. Rediscovery of the giant seastar *Luidia superba* A.H. Clark in the Galapagos Islands. Bulletin of Marine Science, **28**(2):375-376. [*Luidia superba*]

Marcus, E. 1961. Opisthobranch mollusks from California. Veliger, 3 (Suppl. 1):1-85 [*Fiona pinnata*]

Marcus, E. & E.G. Marcus. 1967. American Opisthobranch Mollusks. Part II, Opisthobranchs from the Gulf of California. Studies in Tropical Oceanography, Miami No. 6 (1-2):139-155. [*Tayuva ketos*]

Marshall, J.G. & R.C. Willan. 1999. Nudibranchs of Heron Island, Great Barrier Reef. Leiden, Backhuys Publishers, 257 pp. [*Otinodoris sp.*]

Poorman, L.H. & D.K. Mulliner. 1981. A new species of Crosslandia (Nudibranchia: Dendronotacea) from the Gulf of California. The Nautilus, **95**(2)96-99. [*Crosslandia daedali*]

Valdés A., J. Ortea, C. Avila, and M. Ballesteros. 1996. Review of the genus *Dendrodoris* Ehrenberg, 1831 (Gastropoda, Nudibranchia). Journal of Molluscan Studies, **62**: 1-31. [*Dendrodoris fumata*]

Williams, G.C. & T.M. Gosliner. 1973. A new species of anaspidean opisthobranch from the Gulf of California. The Veliger, **16**: 216-232. [*Phyllaplysia padinae*]

Ziesenhenne, F.C. 1942. New Eastern Pacific sea stars. Allan Hancock Pacific Expeditions, **8**(4):197-224. [*Coronaster marchenus*]

Index

A

Actinostella, 71
Aglaophenia, 127
Aiptasia, 76
alba, Leptogorgia, 111
ampullacea, Sertularella, 131
Anthopleura, 65, 69
Antiparactis, 74
Antipathes, 96
Aplysia, 143
Arachnanthus, 94
Astrangia, 56-57
Aulactinia, 68

B

baumanni, Chromodoris, 144
bellonae, Luidia, 136
benedeni, Botruanthus, 93
benedeni, Cerianthus, 93
Berthella, 143
Botruanthus, 93
bradleyi, Actinostella, 71
bradleyi, Oulangia, 58
browni, Astrangia, 56
Bunodactis, 67
Bunodosoma, 66

C

californica, Berthella, 143
Calliactis, 73
canaliculatus, Tethyaster, 137
capitata, Pocillopora, 14
carneum, Eudendrium, 126
carolynae, Platydoris, 145
Caryophyllia, 61
Cavernulina, 119
Cerianthus, 94

chiriquiensis, Pavona, 28
Chromodoris, 144
Cladopsammia, 46-48
clavus, Pavona, 30
clematis, Phymactis, 71
coccinea, Tubastraea, 51
columbia, Luidia, 136
consagensis, Phyllangia, 60
Coronaster, 138
Crosslandia, 149
Culicia, 55
curvata, Cycloseris, 40
Cycloseris, 40

D

daedali, Crosslandia, 149
damicornis, Pocillopora, 10
dampieri, Pacifigorgia, 112
daniana, Eugorgia, 110
darwini, Cavernulina, 119
darwinii, Pacifigorgia, 113
Dendrodoris, 145
Diaseris, 41
dichotoma, Obelia, 130
diegensis, Aglaophenia, 127
disticha, Pennaria, 125
distorta, Diaseris, 41
Dynamena, 133

E

Ectopleura, 123-124
effusus, Pocillopora, 18
eguchii, Cladopsammia, 46
elegans, Pocillopora, 14
elenalexiae, Favorinus, 148
equatorialis, Astrangia, 57
Euapta, 141
Eudendrium, 126

160

Eugorgia, 110
eydouxi, Pocillopora, 16

F
faulkneri, Tubastraea, 54
Favorinus, 148
Fiona, 148
floreana, Tubastraea, 53
floridana, Plumularia, 134
fruticosa, Muricea, 103
fumata, Dendrodoris, 145

G
galapagensis, Antipathes, 96
galapagensis, Virgularia, 117
Gardineroseris, 36
gigantea, Pavona, 32
godeffroyi, Euapta, 141
gracilis, Cladopsammia, 48
grandis, Bunodosoma, 66

H
Heterogorgia 108-109
hickmani, Heterogorgia, 108
Hormiphora, 99
horrida, Paulia, 139

I
inflata, Pocillopora, 20
integra, Ectopleura, 123
isabela, Polycyathus, 59

K
ketos, Tayuva, 146

L
Leiaster, 140

Leptogorgia, 111
Leptoseris, 38
lighti, Nemalecium, 129
ligulata, Pocillopora, 22
lineolatus, Antiparactis, 74
lobata, Porites, 42
Luidia, 136-136

M
Macrorhynchia, 128
Madracis, 24
Madrepora, 58
maldivensis, Pavona, 34
marchenus, Coronaster, 138
mariscali, Anthopleura, 65
meandrina, Pocillopora, 12
media, Ectopleura, 124
mexicana, Aulactinia, 68
mexicana, Bunodactis, 68
Muricea, 103-107
mutuki, Palythoa, 82
Myriopathes, 97

N
Nemalecium, 129
nigrescens, Anthopleura, 69

O
Obelia, 130
oculata, Madrepora, 58
Otinodoris, 144
Oulangia, 58

P
Pacifigorgia, 112-116
padinae, Phyllaplysia, 142
palmata, Hormiphora, 99
Palythoa, 82-83
panamensis, Myriopathes, 97

panamensis, Telmatactis, 72
papillosa, Phymactis, 70
Parazoanthus, 84-89
parvula, Aplysia, 143
Paulia, 139
Pavona, 26-35
Pennaria, 125
pharensis, Madracis, 24
philippina, Macrorhynchia, 128
Phyllangia, 60
Phyllaplysia, 142
Phymactis, 70
pinnata, Fiona, 148
planulata, Gardineroseris, 36
Platydoris, 145
Plumularia, 134
Pocillopora, 8-23, 25
Polycyathus, 59
polypus, Calliactis, 73
Porites, 42
profundacella, Psammocora, 7
Psammocora, 6-7
Ptilosarcus, 118

Q
quadridentata, Dynamena, 133

R
Rhizopsammia, 49-50
rubescens, Thordisa, 146
rubripunctata, Pacifigorgia, 114

S
sansibaricus, Zoanthus, 80
scabra, Leptoseris, 38
Scytalium, 117
Sertularella, 131
stellata, Culicia, 55
stellata, Psammocora, 6

striatus, Stylocheilus, 142
Stylocheilus, 142
superba, Luidia, 135
superficialis, Psammocora, 7
symbiotica, Pacifigorgia, 115

T
tagusensis, Tubastraea, 52
Tayuva, 146
Telmatactis, 72
teres, Leiaster, 140
Tethyaster, 137
Thordisa, 146
Tridentata, 132
Tubastraea, 51-54
tuberculosa, Palythoa, 83
turbinata, Tridentata, 132

U
undulatus, Ptilosarcus, 118

V
varians, Pavona, 26
verrilli, Rhizopsammia, 50
verrucosa, Heterogorgia, 109
verrucosa, Pocillopora, 8
Virgularia, 117

W
wellingtoni, Rhizopsammia, 49
woodjonesi, Pocillopora, 23

Z
Zoanthus, 80